THE SUBURBAN LAND QUESTION

A Global Survey

Edited by Richard Harris and Ute Lehrer

When discussing the suburbs and the process of suburbanization – namely, the conversion of rural land to urban use – most writers focus on particular countries in the northern hemisphere, implying that patterns and processes elsewhere are fundamentally different. The purpose of *The Suburban Land Question* is to identify the common elements of global suburban development, focusing on issues associated with the scale and pace of rapid urbanization around the world.

The contributors to the volume draw on a variety of sources, including official data, planning documents, newspapers, interviews, photographs, and field observations, to explore the patterns, processes, and planning of suburban land development. Featuring case studies from China, India, Latin America, South Africa, as well as France, Austria, the Netherlands, the United States, and Canada, this volume discusses the peculiarly transitional character of suburban land and addresses the many elements that distinguish land development in urban fringe areas, including economy, social infrastructure, and legality.

(Global Suburbanisms)

RICHARD HARRIS is a professor in the School of Geography and Earth Sciences at McMaster University.

UTE LEHRER is an associate professor in the Faculty of Environmental Studies at York University.

GLOBAL SUBURBANISMS

Series Editor: Roger Keil, York University

Urbanization is at the core of the global economy today. Yet, crucially, suburbanization now dominates 21st-century urban development. This book series is the first to systematically take stock of worldwide developments in suburbanization and suburbanisms today. Drawing on methodological and analytical approaches from political economy, urban political ecology, and social and cultural geography, the series seeks to situate the complex processes of suburbanization as they pose challenges to policymakers, planners, and academics alike.

For a list of the books published in this series see page 341.

EDITED BY RICHARD HARRIS
AND UTE LEHRER

The Suburban Land Question

A Global Survey

UNIVERSITY OF TORONTO PRESS
Toronto Buffalo London

Toronto Buffalo London
www.utorontopress.com
Printed in the U.S.A.

ISBN 978-1-4426-4959-0 (cloth) ISBN 978-1-4426-2695-9 (paper)

Printed on acid-free, 100% post-consumer recycled paper with vegetable-based inks.

Global Suburbanisms

Library and Archives Canada Cataloguing in Publication

The suburban land question : a global survey / edited by Richard Harris and Ute Lehrer.

(Global suburbanisms)
Includes bibliographical references and index.
ISBN 978-1-4426-4959-0 (cloth). – ISBN 978-1-4426-2695-9 (paper)

1. Suburbs – Case studies. 2. Cities and towns – Growth – Case studies. 3. City planning – Case studies. 4. Land subdivision – Case studies. I. Harris, Richard, 1952–, editor II. Lehrer, Ute, 1960–, editor III. Series: Global suburbanisms

HT351.S96 2018 307.74 C2017-908045-8

Collage on chapter opening pages johnwoodcock/iStockphoto and Alhontess/iStockphoto

University of Toronto Press acknowledges the financial assistance to its publishing program of the Canada Council for the Arts and the Ontario Arts Council, an agency of the Government of Ontario.

Funded by the Government of Canada Financé par le gouvernement du Canada

Canada

Contents

Figures

Colour Plates
(following page 206)

Tables

Preface

Many of the papers that are collected here were commissioned for, and first presented at, a workshop on suburban land that was held at the Centre d'Études Politiques de l'Europe Latine, CNRS-University of Monpellier I in the fall of 2012. The editors would like to acknowledge the University of Montpellier for providing assistance, both financial and in-kind. In particular, we would like to thank Emmanuel Négrier for being such a fine host.

The workshop in Montpellier was the second of three that have been held as part of the Global Suburbanisms project, which is headed by Roger Keil at York University, Toronto, and supported by the Social Sciences and Humanities Research Council of Canada. All three workshops address one of the foundational topics of the larger research project: governance, land, and infrastructure. The first workshop on suburban governance was held in Leipzig, Germany, and led to a collection edited by Pierre Hamel and Roger Keil (2015). The third, on infrastructure, took place in Waterloo, Canada, in June 2015, and a publication by Pierre Filion and Roger Keil is forthcoming. We would like to take this opportunity to thank SSHRC for its support and Roger for his continued enthusiasm and intellectual guidance.

Following the workshop, in order to provide wide geographical coverage, we commissioned several additional papers. That is one reason why this project has taken longer to reach fruition than we had originally intended. Another is that it took us some time to fully appreciate the dimensions of the subject. Together with Robin Bloch (and with assistance from Don Brown), we had prepared a document for debate. Translated versions subsequently appeared in French and Chinese (Lehrer, Harris, and Bloch 2015a; 2015b). But workshop discussions, coupled with subsequent reading and conversations, convinced us that a bolder, wider-ranging, and better integrated statement was required. That is what we now try to provide in the introduction and in our concluding suggestions.

This book would not have been possible without the help of the good people at University of Toronto Press and their commitment to the project. We would like to thank Doug Hildebrand for believing in the project and for stick-handling it through the review process until the penultimate stage, at which point Anne Laughlin took over and expeditiously brought the manuscript into the light of day. It was also a pleasure to work with Cathy Frost, who provided us with a thorough copy-editing of the entire book manuscript. And the final touch was put on by Judy Dunlop, whose superb skills produced a masterful index within a blink of an eye. To all, we offer our heartfelt thanks.

Finally, we would like to thank Roza Tchoukaleyska, who has helped us at all stages in the process, providing logistical help in Montpellier and copy-editing skills for the final draft, as well as Loren March, who helped with submitting the manuscript. Without the three Rs – Roza, Robin, and Roger – this volume never would have happened.

Richard Harris and Ute Lehrer
Toronto, August 2017

REFERENCES

Hamel, P., and R. Keil, eds. 2015. *Suburban Governance: A Global View*. Toronto: University of Toronto Press.

Lehrer, U, R. Harris, and R. Bloch. 2015a. "The Suburban Land Question," *Urban Planning International* 30, 6: 18–26. 郊区土地问题 文章编号:1673–9493(2015)06–0018–09中图分类号:TU98;F311文献标识码:A.

Lehrer, U., R. Harris, and R. Bloch. 2015b. "La questione du terrain suburbaine." *Pole Sud* 42: 63–85.

Prologue: A Photo Essay on the Spatial Morphology of Suburban Land Development

UTE LEHRER

The photo essay in the middle of the book draws on John Berger's argument about the relationship between seeing and knowing (1973, 7): "seeing comes before words. It is seeing which establishes our place in the surrounding world; we explain that world with words, but words can never undo the fact that we are surrounded by it. The relation between what we see and what we know is never settled." Therefore, when we analyse processes of urbanization, we also need to look around us, explore the territory, the signs and traces that capital investment leaves in the various places, and, in the context of this book, particularly the outskirts of cities. The photo essay takes up some of the main themes of visual distinctiveness and commonalities of spatially peripheral places that one can observe from the air or the ground (Plates 1a to 8b). The images include highlights from several thousands of pictures I took on my plane travels over the years, images that borrow inspiration from the works of photographers such as Bernd and Hilla Becher (2002), who elevated industrial landscapes into their own right, an art form that is continued through their students from the Düsseldorf Art Academy (Engler 2017); and from Camilo José Vergara's pictures of Detroit (1999), showing the temporality and the other side of growth – namely, decay – of built environment; and finally, Edward Burtinsky's critical contribution of the destruction of our own habitat, or as he calls it, "manufactured landscapes," by processes of urbanization in faraway places (2005). When talking about the suburban land question, one should not forget its visual spatiality in all its articulations and contradictions.

The fascination that the traveller feels by going to places different from the familiar and to learn from them is magnified when she can involve two scales. The first is the view from above. It allows the researcher to see underlying spatial logics and patterns, for example, the impact of railway lines on the urban form; natural conditions such as ravines and rivers; as well as the juxtaposition of different usages, for instance, the footprint of single-family houses versus that of industrial buildings. The bird's-eye view from an airplane best permits such a distant reading of spatial connections and contradictions. While this method of analysis takes on a privileged position, a position that was celebrated within the modernist movement, its critique demands that any scrutiny of urban form also needs a view from below. This close-up engagement goes beyond the visual representation of everyday life. It investigates how human habitats are constructed and function on the ground, with all their messiness.

Processes of suburbanization often entail the transformation of former agricultural land into human habitats. Therefore, two dominant forms of housing, which we can find around the world, frame the essay: the development of single-family homes (Plates 1a, 1b) and the massive housing complex of high-rise buildings (Plates 8a, 8b). Both have been built by private developers, employing the usual mix of financing and building practices, even if some of these developments fail to conform to existing regulations. It is striking that both articulations, the low-rises and the high-rises, follow common geometric patterns, whether as straight or curved lines, despite appearing extremely restrictive for any further development within. Another commonality is the infringement on green space, most often in confrontation with existing agricultural land.

Plates 2a to 3b demonstrate the transitional character of the suburban landscape and the central role of infrastructure and industrial usages. In examples taken from Austria (Plate 2a) and China (Plate 2b), freeways dissect the agricultural land, linking suburban housing developments with workplaces in other parts of the urban area. Meanwhile, agricultural production is still going on. We see single-family homes, apartment buildings, and industrial complexes scattered throughout the landscape. Vast warehouse complexes have become the lifeline for cities within advanced capitalism, providing the urban population with goods for consumption and production, and are usually found in the vicinity of airports and along major freeways. Their rectangular shape and flat roofs (Plate 3a, Munich, Germany, and Plate 3b, São Paulo, Brazil) create an enormous impact on the landscape, not just visually but also ecologically due to their huge footprint, particularly when they are compared with surrounding buildings, as well as global warming as heat islands.

Growth at the periphery of cities can take on different forms, applying diverse ways of building practices. While in Plates 4a and 4b the same building material was used – concrete, which is expensive, needs specific advance planning and is less flexible – they differ enormously in terms of scale, time, capital investment, and procedure. The former example is from Shanghai, China, where entire new cities are built almost overnight on the edge of already existing cities, providing colossal high-rise condominiums but also mid-rise apartments and villas for the rich. In the latter example, we can see an articulation of the incremental process that is used around the world by individuals who will build when money and/or building materials become available. As the example from Greece shows, it is not clear at first if the building has been stopped permanently or if it eventually will be finished. Only time can tell. Very different are the visual snapshots of self-built housing in Gauteng, South Africa (Plate 5b), and that next to a garbage dump in Mexico City (Plate 5a). In both cases people incrementally built their own housing, using whatever materials they could afford or simply found along the way. Despite their small size, usually many people live in these homes.

Since we often assume that urban forms in different cultural contexts should look dissimilar, we are surprised to find out that, in fact, they are often indistinguishable. While this assumption is especially true for cases where developers and government agencies are involved (see above and Plates 1a, 1b, 8a, 8b), it is also true for areas of individual land ownership. Here, we have the examples of Tokyo, Japan (Plate 6a), and Athens, Greece (Plate 6b), where even for a trained reader of aerial photography, it is hard to distinguish between them and, in fact, they share a striking similarity, though they are representatives of two very different and long-standing traditions of building houses. Similarities across different traditions and how the zone of transition towards the green space on the outskirts of today's cities is treated can be seen in Plate 7a (Helsinki, Finland) and Plate 7b (Mexico City, Mexico). Both cases play with the idea of a clear edge, but at the same time they end some of their streets tentacle-like, primed for further incursions into the forested areas, setting the context for more development. This seems to be true for both Finland, where planning processes are highly regulated, as well as for Mexico, where planning policies are secondary to the particular interests of powerful players, be they government or private.

While I am not suggesting that we should generalize from any of those cases, it is the careful engagement with the photographic material of spatial forms that speaks to "the heart of the field" of international comparative research (Robinson 2011, 1). Therefore, apart from their intrinsic appeal, images of the urban landscape can tell us a great deal about development processes at the urban fringe. They are a reflection of the ever-expanding market economy and illustrate the variety of ways in which, as Henri Lefebvre already has forcefully hypothesized half a century ago (Lefebvre 1970, 7), the world has become urban. Consequently, suburban land development is front and centre in the processes of urbanization.

REFERENCES

Becher, B., and H. Becher. 2002. *Industrial Landscapes*. Cambridge, MA: MIT Press.

Berger, J. 1973. *Ways of Seeing*. London: BBC and Penguin Books.

Burtynsky, E. 2005. *Manufactured Landscapes: The Photography of Edward Burtynsky. Ottawa, Ontario, Canada: National Gallery of Canada*. New Haven, London: Yale University Press.

Engler, M., ed. 2017. *Fotografien werden Bilder: Die Becher Klasse*. Munich: Hirmer Verlag.

Lefebvre, H. 1970. *La Révolution urbaine*. Paris: Gallimard.

Robinson, J. 2011. "Cities in a World of Cities: The Comparative Gesture.'" *International Journal of Urban and Regional Research* 35 (1): 1–23. https://doi.org/10.1111/j.1468-2427.2010.00982.x.

Vergara, C.V. 1999. *American Ruins*. New York: Monacelli Press.

THE SUBURBAN LAND QUESTION

A Global Survey

chapter one

The Suburban Land Question: Introduction

RICHARD HARRIS AND UTE LEHRER

"… the suburban situation on the whole is one of transition."

(Douglass 1925, 164)

We, who write about cities, are experiencing a failure of nerve. Never have so many of us written about suburbanization, suburbanism, and suburbs, but never have we expressed so much uneasiness about the terms themselves and about the urban reality to which they are supposed to refer. Urban areas of any size are demonstrably dispersed and polycentric, their outer districts diverse, so diverse that they defy generalization. This is apparent around the globe. It might seem that the urban fringes of Zurich, Denver, and even Halifax, Nova Scotia – all places that are discussed in this collection – cannot easily be fitted into a single mould, and certainly not the hoary stereotype of the residential middle-class enclave (Harris, Lehrer, and Bloch 2013). When we look beyond these places to the periphery of the burgeoning metropoles of South Africa, Latin America, India, and China – also surveyed here – the range of urban forms, incomes, and lifestyles increases again. What can these places possibly share? Increasingly, if only implicitly, the common answer seems to be "not much." Yet the comparative urge, which also underlies this book, is what Jenny Robinson refers to as a "new phase of comparative urban research" where both experimental thinking and foundational rigour are coming together (2011, 1; 2016).

A typical modern approach is adopted in a recent edited collection entitled *Variations of Suburbanism: Approaching a Global Phenomenon.* The editor, Barbara Schönig (2015), notes that old stereotypes, including those that envisaged a neat dichotomy of the city and its dependent suburbs, no longer fit – if indeed they ever did (see Christian Schmid in chapter 4; Harris and Lewis 2001; Soja 1992; Lehrer 1994). Cities, she notes, are diverse in terms of their morphology, their social composition, and modes of governance (Schönig 2015, 8–9). Recognizing this fact, she endorses the suggestion advanced by Michael Ekers, Pierre Hamel, and Roger Keil (2012) that, at its core, suburbanization can be understood as an increase in the population of those parts of the urban area that lie beyond the central city, a trend associated with spatial expansion. This is fair enough as far as it goes, but we believe that there are further and more telling commonalities, of urban forms and even more of underlying processes, and that the key to our understanding of both is the peculiar nature of suburban land. The purpose of our introduction is to put forward this argument by outlining the nature and importance of suburban land as a distinctive issue worldwide. The chapters that follow are designed to illustrate some key aspects of the issue at hand, while the concluding chapter outlines the challenges associated with the research agenda that follows from this point of view. The eight pages of coloured plates in the middle of the book take up some of the main themes of visual distinctiveness and commonalities of spatially peripheral places as one can observe from the air or the ground (plates 1a to 8b).

Echoing Harlan Douglass (1925), whose observation we used as an epigraph, we argue that the unique quality of suburban land is that it is transitional. It may be objected that there is nothing unique about that observation. Cities are always changing, and these days much of the countryside is, too; landscapes and lifeways everywhere are in flux. But suburban land is transitional in particular ways and along four dimensions: geographical, historical, cultural, and above all in the dynamic values that local residents attach to it.

These four dimensions are important in different ways and degrees. Arguably, the most significant is the cultural, if it is understood in its broad anthropological sense, for cultural defines the values associated with land, suburban or otherwise. In western societies, and increasingly everywhere, those values are understood in terms of a binary – use and exchange – where the latter is determined through a more or less regulated market. For that reason, in the second and third sections of this introduction we consider in turn what those values are and what it is that defines the peculiar character of the market for suburban land.

These considerations are very general, equally applicable to London in the late eighteenth century – when the term "suburb" first came into popular use – to Lima, Peru, in the 1950s, and to Shenzhen today. They span what used to be called the First and Third Worlds, then "developed" and "developing," and now the global north and south. But of course, the market for suburban land, and hence the character of the suburbanization process, is driven by different forces in each place and time. London's market was indicated by a rising urban bourgeoisie distributing itself across landed estates; Peru's by massive invasions carried out by rural-urban migrants; Shenzhen's by migration, too, but also by globalized capital markets and supply chains. It awould be an interesting task, and a valuable service, to trace the changing mix of historical forces that have shaped the diverse historical geographies of suburban land markets. That is beyond our ability or purpose. But in the latter parts of this introduction we sketch the main elements that define the current era and then offer a sort of checklist of issues that future research needs to consider, whether it involves a local case study or a wider national, or indeed international, survey. But we want to start with what we mean by the dynamic values that are attached to the geographical, historical, and cultural dimensions of suburban land.

The Transitional Character of the Suburbs

Suburbs are transitional in space, in time, and in their cultural aspects. The geographical and historical transitions are apparent in the landscape or on a map; they pertain above all to urban form. The cultural aspects, understood to involve a whole way of life, include the processes, often invisible and intangible, that shape landscape transitions. They have economic and political as well as social aspects and are focused, most importantly, through the market for suburban land.

Space and Time

The intermediate quality of suburban space is obvious to anyone who has driven, ridden, or flown into a major metropolitan area: it lies between city and country, the urban and the rural. By "urban" we refer to territory that is fully built up at relatively high densities; "rural" consists of a variable mix of agriculture, wasteland, and wilderness. In the past, when many urban areas were strongly monocentric, suburbs occupied a zone that was broadly circular or, in the case of ports or other cities with major natural hindrances, semi-circular. Today, patterns are

Figure 1.1 Juxtaposition of different land uses. Photo: Ute Lehrer

typically messier: somewhat so around small and medium-sized centres such as Hamilton, Canada, which has been Richard's home town since 1988; more so around Toronto, Canada, or Zurich, Switzerland, Ute's current and former base, respectively (we are acutely aware that our way of thinking as it is represented in these notes is a result of our own experiences, of where we have lived and what we have studied in greater detail); and even more so in many US metropolitan areas where local government is more fragmented and where land-use regulations are even less strict. Indeed, as Wolfgang Andexlinger, Pia Kronberger-Nabielek, and Kersten Nabielek show in their comparative piece on Austria and the Netherlands in chapter 7, it is a misconception to think that planning and governance in Europe are integrated in an adequate way.

But even where there is discontinuous sprawl, such that the boundary between urban and non-urban becomes blurred, a boundary still exists. As James Fallows (2016) recently observed, from the air especially it is still very apparent (see also Ute Lehrer's colour plates of aerial photography in the midsection of this book). And it can be meaningfully mapped, as Shlomo Angel and his associates have done quite recently in their *Atlas of Urban Expansion* (Angel et al. 2012). There is, in fact, a broad urban periphery: it extends from urban peripheral areas that are fully developed into an outer exurban or peri-urban fringe where urban and rural land uses intermingle. Some would describe the inner part as suburb and the outer as sprawl, but for present purpose we subsume both under the "suburban" label. Beyond it is a territory we can call the countryside, even though in places like Vancouver or Cairo it may be mountain or desert. Suburban areas, then, are places in between.

To the visitor, one of the most striking, visible features of this transitional territory is that land here is usually used more liberally than in the city's core. In North America, Britain, and Australia, as indeed in many other parts of the world, the obvious indicator is that the city's high- and mid-rises give way to townhouses and then single-detached houses on increasingly large lots (see Plates 1a, 1b, 6a, and 6b). Elsewhere, in parts of Europe, Latin America, the Middle East, and China, we find high-rise buildings in the periphery of urban areas, sometimes in the form of towers in the park and other times as a highly concentrated settlement next to single-family homes or agricultural lands (see Plates 8a and 8b). One of the consequences is that population and building densities, which are usually measured by the ratio between number of floors and built-up land, away from the urban centre tend to decline (but as we know, this ratio does not tell the full story, particularly when it comes to high density, which we find in some of the self-built and often informal settlements around the world). As many researchers, including Angel and his associates (2012), have shown, such density gradients were long a feature of cities in the global north in the nineteenth and twentieth centuries. They still characterize rapidly growing metropolises in the global south, such as Beijing, as well as smaller cities such as Bogotá and Cali, Colombia (Mohan 1994, 58; Zheng and Kahn 2008). For several decades the major exceptions were cities in the Soviet bloc as well as several continental European countries, including France, Germany, and Italy, where we find a concentration of high-rise buildings outside the city centre. As Sonia Hirt's study of Sofia in chapter 2 shows, post-war high-rise apartment buildings erected by the state produced densities that sometimes increased towards the fringe. Here, however, the introduction of land markets from the 1990s onward immediately began to challenge that pattern (Bertaud and Renaud 1997).

Another feature that is apparent to anyone in transit to or from the centre is that suburban territory contains a lot of visible infrastructure (Young, Wood, and Keil 2011), including self-storage facilities, electricity transmission towers, and pipelines, but it pertains overwhelmingly to transportation of people and goods: highways, parking lots, rail lines, freight yards, truck terminals, warehouses, storage tanks, and airports. Much of this infrastructure serves the city, linking it to the region and places beyond; it is, precisely, what visitors, commuters, and truckers need in order to get to and from the centre. But when such routes approach the core, they often disappear underground, adding a hidden dimension of density. Other infrastructures serve the whole metropolitan area but are simply too land-extensive to be fitted into the city, so they help define the urban periphery as a place where land uses associated with movement occupy a lot of space (see Plates 3a and 3b). Interestingly, this fact is at odds with the stereotype of the suburbs as being, quintessentially, residential in character (Keil 2017).

Much less obvious, except to the long-term resident, suburban areas are also transitional spaces in historical terms (McManus and Ethington 2007). The city and the country may change, but in most cases their overall layout has been that way for many decades, perhaps centuries, or even millennia. In contrast, a suburban area is temporary. This feature is most obviously true of the outer, peri-urban territory (Fureseth and Lapping 1999; Ginsburg 1991; Simon 2008). Even a casual visitor can sense that a landscape that juxtaposes piggeries, junkyards, cornfields, motels, and scattered housing, which surrounds cities from Canada to France to China, cannot resist the growth pressures of capitalism indefinitely. On a longer timeline, the same is true of the spanking new subdivision. Of course, its basic pattern of land use may not change for decades, even as trees and gardens mature, because, as Pierre Filion discusses in chapter 5, land-use regulations tend to keep things locked in for the long run. On individually owned lots property owners are likely to resist change, especially if they believe that their property values may be affected adversely. As Jill Grant shows in chapter 8 for Halifax, Nova Scotia, such property owners most probably succeed in smaller cities where land values are lower, the pressures for redevelopment weaker, and local expectations about neighbourhood change are different from those in places such as Markham, Ontario, in Toronto's outer suburbs. But, as long as the urban area continues to grow, as most do, the relative position of this subdivision will move further and further from the fringe until it becomes, in effect, an urban neighbourhood. This is the evolutionary course followed by many of the districts, from Greenwich Village through Harlem to Brooklyn's Park Slope, that we now think of as unambiguously urban. The evolution began in older metros, and so, disproportionately in the global north, but in light of the rapid rates of urbanization that have characterized the period since 1945, it is now apparent everywhere. For example, Evers and Korff (2000, 221) have observed the transformation "from village to suburb to finally city quarter" on the periphery of Bangkok. Eventually, almost every suburban subdivision will come under pressure, from the market and/or the state, to densify. Residents may resist, but even if only belatedly, their area will almost certainly grow denser, faster paced, and in those respects more urban.

The Culture of Suburbanism

Enlivening these geographical and historical transitions is one of culture. Numerous words have been expended in trying to characterize the suburban way of life, and in critiquing the idea that there is such a thing. Some writers have associated suburbanism with familism, as the habitus of families with children who seek greater privacy and outdoor play spaces, whether public or private, together with a safe removal from the adult temptations and perceived risks of the city (Clapson 2003; R. Harris 2004; Jackson 1985; Gans 1969), a perspective that was heavily put forward by the sales strategy employed by developers of the suburbs (Keil and Graham 1998). In many places and times there may have been good grounds for those perceptions. In those places, and many others besides, suburban living normally mandates the use of privatized modes of transportation: public transit is less viable, meaning that it is less frequent, less convenient, or simply absent. This reflects not only the simple economic arithmetic of lower residential densities of most places but also the complex geometry of metropolitan living as well as the political will for infrastructure investment. Whatever their density, places that are less central are, on average, further from other places in the metropolitan area and for that reason are less accessible.

The weight of historical inertia confirms the contrast. Most fixed-rail transportation routes were built to connect the city with its suburbs, and the high cost of supplementing radial with circular routes, coupled with problems of securing agreement from relevant parties, means that access to suburban transit lags behind even the levels that lower densities would suggest. The fact that expensive encircling projects such as London's Crossrail or Moscow's projected Third Interchange Contour are so exceptional serves to underline the point. As a result, and given the distances that suburban residents often must travel to work or to shop, "private" usually means a car, or possibly a moped, and sometimes a bicycle. As Pierre Filion carefully demonstrates in chapter 5, assumptions about auto use frame the design of modern suburbs at all scales. At the same time, automobility shapes everyday life in the suburbs. It embodies a lifestyle that arguably also brings its own, more privatized, politics (Walks 2008, 2013).

Automobility may distinguish the suburban subdivision, and even more the peri-urban fringe, from the city neighbourhood, but not from the country. There are still parts of the world where peri-urban living goes with horse- or bullock-drawn carts, not to mention bicycles and long walks, but in most places some version of the internal combustion engine (or, increasingly, the electric motor) is the key enabler. Here, however, a more profound cultural difference comes into play, one with wider if diffuse ramifications (Douglas 2006; Goddard 2012; Savage and Lapping 2003).

Life in a rural setting imposes particular, above all seasonal, rhythms on life. Of course, these rhythms can vary. They commonly involve the busyness of planting in the spring and of harvesting in late summer and fall. In warmer climates the seasonal rhythms may be framed by variations in rainfall rather than temperature and accommodate more than one harvest. Rural life even establishes patterns for part-time hobby farmers and back-to-the-landers who like to grow their own vegetables. It is more subtly true for those exurbanites who unavoidably experience the rural landscape, especially if they have made an aesthetic choice. The changing natural scene may be no more than a daily backdrop, but it is unmissable and, psychologists tell us, has effects whether we are aware of them or not (Capaldi, Dopko, and Zelenski 2014). In contrast, urban living is ruled by the clock: the nine-to-five, the planned meeting, the transit schedule, or, increasingly, the challenges of juggling two or more precarious, part-time jobs in different locations. These factors characterize a way of life in which measurement of all sorts becomes precise: of money above all, but also of living space. An elaborate superstructure of argument could be built upon this contrast, one that goes beyond lifestyles to implicate values. In the past many writers – in a long tradition that in the United States is most closely associated with Thoreau – have coloured this contrast with a romantic glow. The point can be overstated and too easily taps into personal prejudices, but a significant contrast does exist.

Most suburbanites stand unambiguously on the urban side of this divide. A century ago in Britain and North America some of their advocates spoke of the suburbs as places that married city and country (Howard 1898; Clapson 2003). They thought in terms not only of residential densities and significant amounts of greenery but also of personal values. It is likely that they always overstated, or indeed misstated, the case. Today, at any rate, many people choose suburban living not for any supposed quasi-rural lifestyle but simply because suburban homes are cheaper, if paid for in part by longer commutes. This is true for Toronto (see R. Harris 2015), as it is for Denver and Montpellier, as Françoise Jarrige, Emmanuel Négrier, and Marc Smyrl show in chapter 10. Suburban lives are fully ruled by the clock, none more than those of adults who have children, whose daily routines must accommodate school schedules and extracurricular activities as well as the demands of work.

But the life routines of some suburban residents are not so neatly categorized and may tip the other way. There are those arrivals who make an aesthetic choice about suburban, especially peri-urban, living. They can play a non-trivial role in suburban politics by resisting further suburban development in the name of preserving landscapes, wetlands, and watersheds (see Sandberg, Wekerle, and Gilbert 2013). And then there are those who have been unwittingly, and often unwillingly, drawn into the city's ambit. They include farmers, of course, and also those who serve them: truck and equipment dealers, seed merchants, and the range of businesses that are needed to support any community, from restaurants to places of worship. The tensions between these long-standing rural interests and the incomers can be as fraught as those between both groups and the agents of the expanding city. If there is a suburban marriage here, they are nuptials forced by fate or, if you prefer, by large and inscrutable forces, not entered into voluntarily on the basis of shared values.

The Values Attached to Suburban Land

It is, above all, the dynamic juxtaposition of values that most clearly sets the suburban environments apart and eventually shapes the nature of their geographical and historical transitions. Many of these values pertain to the meaning of land, but before we consider them we need to pause and contemplate: what is meant by "value"?

We routinely distinguish two types of value, in use and in exchange, recognizing that there is a relationship between the two. Rural land differs in kind from urban in that both types of value depend on its intrinsic quality: fertility, determined in part by climate, and its capacity to provide raw materials, including those used for building: stone, aggregate, bricks, and timber. Certain types of urban use also can be attracted or repelled by the quality of particular sites: high-rise developers usually avoid unstable subsoil; builders of exclusive homes are drawn to views and particular landscapes. The founders of cities have generally shunned swamps, vulnerable locations, and deserts. There are, of course, exceptions, Venice and Dubai being examples, and the general rise in sea levels promises to redefine vulnerability for many coastal cities that were once safe. But historically these places have been unusual, although unfavourable conditions increasingly are overcome with technological advancements, political will, or economic power (see, e.g., Ordos City in Inner Mongolia). Most sites can support almost any type of urban use. Their usefulness depends on their location in relation to other sites: how close to a supplier or customers or the market; how convenient to work, shops, and schools. For businesses, it is a question of how profitable a location is and therefore what site rent it can afford to pay. For residents, affordability involves above all a trade-off between the costs of transportation and housing, where variations in the latter are determined by the price of land. The geometrical calculations here are not purely Euclidean, for it is time-distance that matters. Transportation routes warp the surfaces of useful accessibility in ways that all city-dwellers know, and often only too well (see Plates 2a and 2b).

Commercial values embody these contrasts. The price of rural land depends primarily on its intrinsic quality – what it can yield – whereas the price of a parcel of urban land depends, in that hoary saying, on three things: location, location, and ... well, you know the third. Location matters for rural sites, too, especially where bulky agricultural produce or a building material is being shipped to markets, a fact that the nineteenth-century German economist von Thünen (Johnston 2000) knew very well and that was later expanded into the Central Place Theory of Walter Christaller (1933). But, given the ubiquity and much reduced cost of rail, truck, and air transportation, the relevant scale is now one of scores or even thousands of kilometres. The value of urban land can vary significantly over far shorter distances. The length of the working day sets limits on how much time anyone can spend commuting and, depending on the available means of transportation, for most people this indicates that sites more than two hours from the workplace are effectively valueless to the commuter. In general, the more accessible the site, the more valuable, which accounts for the fact that land values almost always decline away from a major urban centre except for subsidiary peaks around suburban sub-centres.

However, today's urban agglomerations are not the same kind as the one that provided the basis for the studies by von Thünen, Christaller, or the Chicago School, where the city was the main unit of analysis. While suburban land development historically was linked to property ownership, industrialization, and displacement, the current period is driven by "post-Fordist regional economies, globalization and neoliberalization" (Keil 2018, 177). Ever since economic restructuring and globalization produced new forms of urban areas (Soja 1992; Lehrer 1994, 2013), the underlying logic of land value is no longer as simple as it was in the nineteenth and early twentieth centuries – if it ever was. Urban areas have become polycentric (Hall and Pain 2006) and are "vast assemblages of human settlements" (Friedmann 2014, 552), bound together through "complex, distant and less widely understood networks" (Gandy 2012, 123).

This new reality questions our conceptual thinking about what constitutes the urban and, with it, the attached value (Keil 2013). Already in 1970 Henri Lefebvre (2003) declared that most of the earth was urbanized, and hence he called this process planetary urbanization. Based on Lefebvre's argument, current literature is proposing not to understand the urban and suburban as bounded entities but instead to see them being exposed to, and integrated into, the same processes of extended urbanization (Brenner and Schmid 2014; Keil 2018). Nevertheless, it is plausible and justifiable to look at suburban land production empirically and conceptually. From the planetary literature we know that land production and the question of land must be seen within a worldwide context, described by Lefebvre as implosion and explosion, and that suburbanization is embedded in processes of extensive urbanization (e.g., Sevilla-Buitrago 2014; Ajl 2014; Angelo 2017). The issue concerns taking possession of land, with a relationship between primitive accumulations and urban commons (Arboleda 2015). Land taking, original accumulation, and accumulation by dispossession are front and centre in the suburban land question (AlShehabi and Suroor 2016; Goonewardena 2014), just as financialization is (Kaika and Ruggiero 2016; Haila 2016; Le Goix 2016; Christophers 2017; Shen and Wu 2017).

This way of theorizing about the urban has been central to Christian Schmid's own work (2005, 2006, 2012; chapter 4 of this book) and more recently in his collaboration with Neil Brenner (2013, 2014; Brenner and Schmid 2014, 2015), where they invite us to rethink the processes of urbanization: "This emergent planetary formation of urbanization is deeply uneven and variegated, and emergent patterns and pathways of sociospatial differentiation within and across this worldwide urban fabric surely require sustained investigation at various geographic scales" (Brenner and Schmid 2014, 747–8). Similar ways of thinking were expressed in the 1990s by Roberto Monte-Mór, who stated in the context of the Brazilian Amazon that "urban-industrial processes impose themselves over virtually all social space" (2005, 942). These can be faraway places, such as the mining industries in the Chilean Andes (Arboleda 2015), the sorts of new development dubbed "Edge Cities" by Joel Garreau (1991); in-between spaces as described by Sieverts (2002) with his term of "Zwischenstadt"; or older settlements that have been encompassed by the growth of a larger and more dynamic place. Zurich presents an interesting example: while constrained by the surrounding topography, it is typical in explaining the relationship between economic restructuring and urban form at the end of the twentieth century, when cities no longer can be described by only the concentric model. Ute Lehrer (1994) used it as the basis for an argument about the growing importance of what she called "FlexSpace," highlighting the relationship between economic and structural changes within a system of flexible accumulation and its spatial articulation. In chapter 4 of this book, Christian Schmid builds on this argument and invites us to rethink categories such as "urban," "suburban," and "rural" in order to comprehend urbanization in the context of Zürich, Switzerland; by doing so, he proposes a territorial approach for urban analysis in general. Almost every major urban area provides one or more illustrations of the complexity of these new urban configurations. This is perhaps especially true in southeast Asia, where the high density of rural settlement has meant that innumerable villages have been incorporated more or less effectively into expanding metropolitan areas (see chapter 12 on Chennai).

The overall price gradient was documented most thoroughly for Chicago in the nineteenth and early twentieth centuries by Homer Hoyt (1933). Despite higher suburban densities, it soon asserted itself in Russian cities after property trading was reintroduced in 1992 (Bertaud and Renaud 1997). Today it characterizes almost all urban areas, varying in size and character from Bamako, Mali, to Bogotá and Cali, Colombia, to Nairobi (Durand-Lasserve, Durand-Lasserve, and Selod 2015; Kimani 1972; Mohan 1994). The existence of this price gradient explains why suburban building and population densities are intermediate between those of the city and those of the country and, dynamically, why the morphology of suburbs is always in transition.

But geographically and historically, changes in price and density gradients are not smooth. This fact matters because the value of land under urban use is always greater, usually much greater, than under any type of rural activity; the difference is commonly an order of magnitude. This is a staple fact of any introductory text in urban land economics. For example, Tony Crook, John Henneberry, and Christine Whitehead (2016) use the conversion of agricultural to residential land use as their prime example of the dramatic change in land values that can come with a change in land use. Examples may be found everywhere. In the first half of the seventeenth century, when Amsterdam was experiencing a real estate boom based on colonial trade, Frans Oetgens became the city's de facto planner (Shorto 2013, 199). He persuaded the city to abandon its policy of prohibiting development outside the walls; meanwhile he was buying suburban property that appreciated by 800 per cent in four years. Today, land zoned for rural use just outside Vancouver is worth barely a tenth of its urban potential, a fact that some owners are using to their advantage while they deceive assessors about its current use (Tomlinson 2016). This is an example where regulations that are introduced in order to protect usages are being exploited by "economically well-situated" people; in a certain way it is reminiscent of a different version of informality that is conceptualized in Ananya Roy's work (2005, 2009, 2011, 2015). Similarly, large differences in land value have motivated the developers of peri-urban sites outside Chennai, India, that Bhuvaneswari Raman describes in chapter 12.

For present purposes, an especially interesting graphical representation was provided by Harold Dunkerley (1983) in a classic survey of urban land policy around the world (Figure 1.2). As the horizontal axis indicates, Dunkerley aimed to show how the value of a parcel of land changed over a period of four decades as it moved from rural to what he calls its "first urban use," most likely as a suburban subdivision, and then to its "second urban use," as the site was redeveloped at a higher density. Each of the two land-use transitions brought a sharp jump in market value, albeit smoothed somewhat by a short anticipatory period (cf. Brown, Phillips, and Roberts 1981). The time frame is unrealistically compressed: most suburban sites are not redeveloped after only twenty years. But the core idea is valid, incidentally illustrating the transitional character of suburban land.

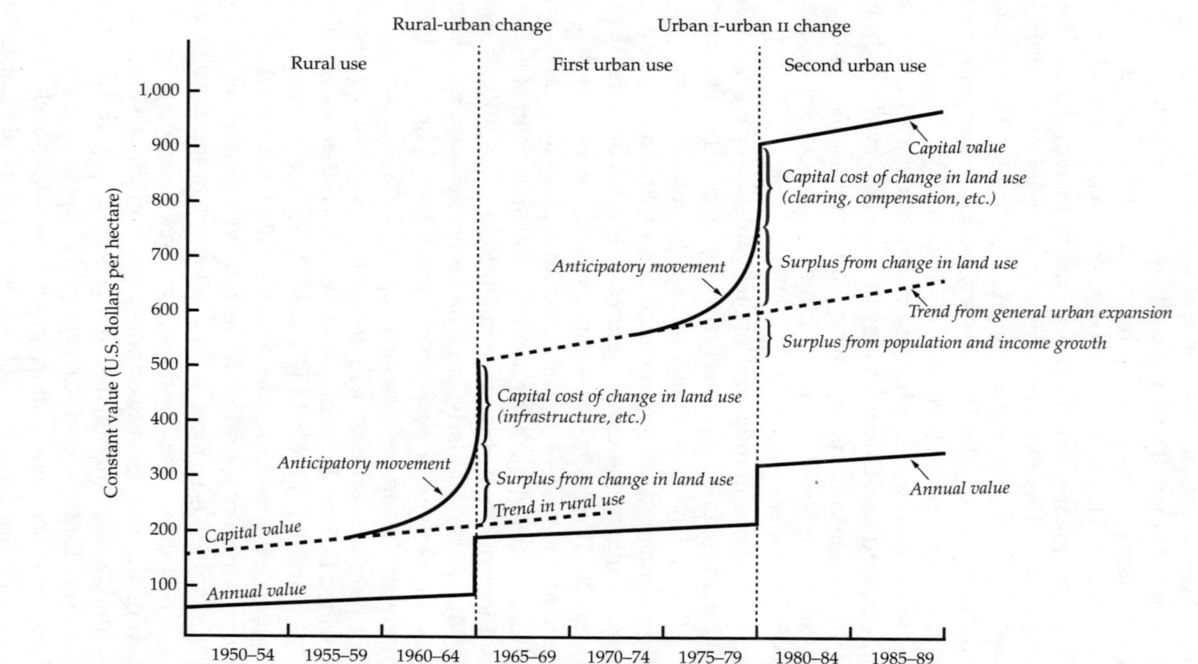

Note: In this illustrative example, capital value is three times annual value under stable conditions. Value changes owing to general inflationary trends are excluded.

Figure 1.2 Changes in annual capital values of land with changes in land use. Source: Dunkerley (1983). © World Bank

It is because land jumps in price when it is converted to urban use that owners of urban land attach much greater significance to the precise measurement of boundaries than do their rural counterparts. A farmer is not likely to care greatly if a neighbour encroaches two feet onto his land; a homeowner will complain and, if necessary, sue. If time is money, so is space. The prospect of a jump in price offers existing owners of rural land at the urban fringe a large incentive to convert to urban use, or to sell at a good price to someone who will. It also invites speculators, those who are on the lookout for the anticipatory movement shown in Figure 1.2.

Given that land is expensive and often bought on credit, price jumps also catch the eye of financial institutions, including banks and pension funds, that seek profitable investments. In the past thirty years or so this has become doubly true with the globalization of capital markets. Suburban land, then, is dynamically transitional, most obviously in land use, but also in the way that it is valued – intrinsically or by location – and inevitably also in price.

Most North Americans and to a lesser extent Europeans are accustomed to this way of thinking: that land is a commodity whose price is determined by its utility, and that it will be used in a way that is most useful, or profitable. But that is not the way that land has always been valued – not in Britain or continental Europe or, certainly, in the white-settler colonies where fee simple land tenure now rules. This issue has been viewed synoptically, among others by John Weaver in his history of the way the British, in particular, exported and over a period of centuries imposed variations on a particular understanding of property rights (2003). It can also be seen at the microscopic scale and in specific suburbs. For example, Ronald Karr (2015) has shown how, between 1770 and 1850, a farm on the outskirts of Brookline, Massachusetts, was made suburban. The progression was accomplished slowly by a nascent group of lawyers, developers, and real estate professionals who not only converted land to a new use but in the process made it into a commodity.

Beneath this process were acts of appropriation. In North America, Native Peoples had a very different understanding of land from that of European colonizers. A typical example was – and still is – the Musqueam band on the west coast of Canada. They recognized a limited form of private property, one that could be inherited by the eldest son, though he was required to share the use of this land with all family members, while most lands in Musqueam territory were available to all. When signing treaties with the British, band members were not aware that they were giving away anything other than shared use (Weightman 1972). Through force, by 1916 the Musqueam were confined to a reserve located at the exurban fringe of Vancouver in a region that became suburban in the early post-war decades and is now part of the central city. For the past century they have experienced an enforced hybrid arrangement. In western fashion they have exclusive rights to their territory, but are free to dispose of their land using modified customary practices. Slowly but steadily they have adopted western lifeways. An early shift was towards settled agriculture and wage labour (C. Harris 1992, 59–60). Later, given their exurban then suburban locations, the Musqueam became the first Native band in Canada to enter into western-style real estate development (Weightman 1972, 35). Porter and Barry (2016) speak about cultural "context zones" in relation to planning and indigenous land rights.

Even in Britain, the locus of much thinking and law regarding the land market, the meaning of land as a commodity varied and has evolved with shifting forms of tenure. This was strikingly true over a period of centuries, but could be apparent in specific places over much shorter periods. For example, Jane Springett has shown that, around suburban Huddersfield in the mid-nineteenth century, some land was held in freehold, some on tenancy-at-will (very insecure), and some in the forms of leasehold on 99- and then on 999-year leases (1982, 26–7). Indeed, Fitzgibbon (2011) has argued that a market for land and housing did not emerge in Britain until the nineteenth century. Today, in Anglo-American law alone there are more than fifty forms of tenure, each with a distinctive bundle of rights (Doebele 1983, 70–1).

In one respect, recent developments have complicated rather than simplified this picture. Initially in parts of Europe and then the United States but now around the world, a partial version of communal tenure has become widespread. Common Interest Developments, many of them gated, have become popular because they offer prestige and the promise of security (Cribbet 1963; McKenzie 1994, 2011; Sparkes 2003; Glasze, Webster, and Frantz 2006; Lehrer and Wieditz 2009; D. Harris 2011; Lehrer 2012, 2016). They are based on condominium tenure, in which certain property rights are private, while other are shared. In chapter 3 of this book André Sorensen shows that this form of tenure may have stored up some challenges for future redevelopment that few have thought through. Tenure is never a simple matter.

And the logic of private property is not the whole story. Even today, non-Indigenous owners often treat urban land as more than just real estate, an instrumental commodity to be devoted to its "highest and best" – that is, most profitable – use. Alan Evans (2004a, 248), in his text on urban land economics, points out that the behaviour of the owners of land depends on their attitudes, attachments, expectations, and goals. Brown and Roberts (1978, 4) provide an example. Around Phoenix, Arizona, they found that the willingness of farmers to sell to developers, even at a substantial profit, depended on their age. In an urban context homeowners and resident landlords, too, develop attachments that often override economic considerations, at least in the short run. Most of us are familiar with situations where a redevelopment was stymied, or shaped, by one or more owners who simply did not want to sell, at any price. And, in one way or another, a good deal of land is owned by the state for use as streets and highways, parks, public spaces, or government buildings. Even dyed-in-the-wool advocates of the market concede that we need such publicly owned spaces. Nowhere is all urban land a commodity, or expected to be.

Some societies attach unique and inalienable values to particular sites, whose ownership and use may not be changed. Others treat ownership as communal, so that decisions about use and possible sale, even when allowed, become complex. Krueckeberg (1999) and especially Bruce (1988) have argued that almost all forms of land tenure make provision for both individual and communal rights, but the balance varies. Some extend that communal element to past and future generations, making the very idea of changing land use moot. Simpson (1976, 224) repeats the apocryphal statement of an African chief that in his culture "land belongs to a vast family of which many are dead, few are living, and countless numbers are still unborn." This practice can be found around the globe, including in Switzerland and Germany, where in certain municipalities the communal forest is understood as being in collective ownership, and where every citizen of the municipality has the right to make use of excessive wood, a practice that has been passed on from generation to generation over centuries.

That is why many observers recognize the coexistence of two or more forms of land tenure. This is a type of legal pluralism that further complicates the valuation of land and of the ways in which local residents decide how land is to be used (Santos 1977; Tamanha 2000). Given the extent of local diversity and the paucity of systematic evidence, it is impossible to generalize about the proportion of urban land in these regions that is treated as a commodity, or to sketch how that land is articulated with other forms of tenure. One of the most perceptive observers of land markets in the global south, Alain Durand-Lasserve, has suggested that we think of this complex articulation as the local "land delivery system." In Bamako, Mali, he and his colleagues have discerned three elements: the private, formal mechanisms that we often refer to as "the market"; the public or parapublic; and the "customary" (Durand-Lasserve, Durand-Lasserve, and Selod 2015; see Evers and Korff 2000, 183; Jenkins 2002, 212; 2009). Even more than "private," "customary" embraces a wide range of practices. What they have in common is that the distinction between public and private, which intuitively seems obvious to most North Americans, is blurred or even meaningless. This triad may serve as a useful approximation of the way suburban land is assigned in many other African and Asian urban areas.

For present purposes, the fact that land can be valued in different ways is important because here, too, suburban areas are commonly the key zones of transition. Almost everywhere, even in Russia, China, and now Cuba, some version of private fee simple land tenure based on capitalist principles is associated above all with cities. Higher prices encourage a more careful measurement, documentation, and enforcement of property boundaries. That is why, very soon after Hong Kong was founded by the British, a precise land survey and system of record-keeping was introduced; much later, as land values soared and technology improved, a re-survey was undertaken to higher standards of precision (Nissim 2008, 137). A similar level of precision, coupled with a rigorous elimination of tenurial ambiguity, is apparent in Singapore, another high-density city-state (Haila 2016, 219). There are economic and historical reasons for this meticulousness. In economic terms cities are dynamic environments where the advantages and pressures of land-use change are exceptionally strong. Except for public spaces that are deemed to have special significance or utility, the persistence of sites whose use is frozen, or of methods of decision-making that are cumbersome, is likely to be viewed as dysfunctional and, sooner or later, will be shunted aside. The same can happen in rural areas, but even more slowly, because the pressures are usually not felt to the same degree.

Then, too, historically large parts of the globe – extending from the Americas through sub-Saharan Africa to Australasia and south, as well as much of southeast Asia – were colonized by one or other of the European nations. Eventually, led by Britain, these powers exported legal systems that included new forms of land tenure, thereby creating at the minimum a form of "tenurial dualism" (Bruce 1988, 31). They were most eager to establish these systems in the port cities, where most Europeans lived and that were crucial to the operation of a colonial mercantile economy. That legacy still frames land ownership and land-use decisions in many cities in the global south, but less so in rural areas.

In these contexts urban growth and suburbanization brings a new discourse about land into rural territory. That is what happened around Brookline, Massachusetts, and Huddersfield, England, in the eighteenth and nineteenth centuries and around Beijing and Bamako in the twentieth and twenty-first centuries. Today, farmers near Toronto resist the encroachment of developers' subdivisions as a threat to their way of life, but in time they may be mollified by generous compensation. Indeed, many have come to anticipate and expect that their land will appreciate in value, and so they resent attempts by governments to curb sprawl by freezing development. They were part of the public debate in Ontario surrounding the province's decision to establish a green belt around Toronto and other urban centres (Macdonald and Keil 2012; Sandberg, Wekerle, and Gilbert 2013).

Around many cities it is a different story. The local equivalents of motels and junkyards are not just visible signs of probable changes in land use and value, but symbols of new ways of thinking and being. Here, the transitional character of suburban land is the thin end of a very large cultural wedge. Six Nations, one of the larger Native reserves in Canada, lies at the edge of the exurban commuter zone of Hamilton, Ontario. In recent years armed conflicts have arisen over the development of disputed land in Caledonia, a border town ("Grand River Land Dispute" 2016). Such disagreements are dwarfed by equivalent disagreements at the fringes of urban centres in large parts of Latin America, Africa, and southeast Asia. To make sense of its significance and dynamic, we need to look more closely at the nature of land as a commodity, and at the markets through which it is traded and its use is determined.

The Suburban Land Market

Although some economists and several politicians have spoken as though it is possible to imagine a pure, perfectly free market, no such thing is possible. Regulation of some sort, typically by the state, is necessary, if only to ensure that contracts are enforced (Block and Evans

Figure 1.3 Fighting for their right to stay. Self-built housing on the fringe of Mexico City, Mexico. Photo: Ute Lehrer, 2014

2005). At the same time, the particular character of any market is shaped by the cultural context and, perhaps even more obviously, by the nature of the commodity in question.

Urban Land Markets

The market in land is so distinctive that in at least one fundamental respect it barely qualifies for the label at all. Part of the usual definition of a commodity is that it be fungible, that is, mobile and substitutable. In contrast, land is fixed in place, with a variety of consequences (Andersson and Moroni 2014; Bertaud 2010; Doebele 1983; Evans 2004b). It is never traded – only the rights to its use are. Even on the smallest scale, and however similar, no two sites are exactly alike; a site in one neighbourhood is never equivalent to one in another. On the urban scale, land markets are notoriously local, so that booms in one can coincide with busts in another. For most purposes it is impossible to speak about a global, or even a national, market in land in the way that one can for oil, automobiles, or coffee.

One of the most important consequences of land's immobility is that every site is unavoidably associated with externalities. These are the costs or benefits that are produced by one person and felt by another but that are unpriced and uncompensated. The stereotypical example is that of pollution produced by a factory and experienced downwind. The cost, in comfort and health, can be considerable, but will not be compensated for (or reduced), unless the polluter is compelled to do so, usually by government. Almost all externalities have a strong geographical component, so that sites adjacent to the source experience much more substantial costs (or benefits) than those more distant, noise being a prime example.

Together, the immobility of land and the ubiquity of localized externalities do much to define the nature of the land market. Urban densities compel regulation – of noise, pollution, traffic, views, and, by extension, land use. That is why such regulations were introduced first in cities, and why many exist only in cities. Noting that such regulations are much of what urban governments do, Jacobs and Paulsen (2009, 135) have observed that "planning is fundamentally about the allocation, distribution and alteration of property rights." Regulation, then, is as much a uniquely urban phenomenon as density is. Construction by amateurs, for example, is often viewed as unexceptionable, or even necessary, in rural contexts, but may be effectively prohibited in the city because of perceived fire or health hazards. Building standards in cities are higher. The Chinese villages that urban expansion has surrounded over the past forty years stand out as some of the most striking examples, as many authors have discussed, including Fulong Wu and Zhigang Li in chapter 6. For that reason, if informality is understood as the evasion of government statutes (Roy 2015), by-laws, or taxes, then it too may be thought of as peculiarly urban (R. Harris 2016; Portes and Haller 2005). At the very least, its potential extent is greater.

Governments also come under pressure to provide more public infrastructure. Private wells, privy pits, and septic tanks can serve farmers and villagers but create serious problems in denser settings. Water and sewer pipes become necessary and typically require funds and regional planning that are beyond the reach of private developers. Similar requirements push governments to plan and provide the transportation infrastructure that makes it possible for the city to function. High densities and limited access to open space create a public demand for parks. Cities, then, require more, and a wider range, of public infrastructure, even to the extent of entirely removing some sites from the market.

If the urban context encourages governments to constrain the land market, it also force-feeds it. In many regions a good deal of farming is for subsistence; even where commercial agriculture reigns, land use may not change for decades or even centuries. In contrast, cities and patterns of urban land use are restless (Knox 1991), encouraging and often necessitating frequent and rapid changes in ownership. They are places where privatized forms of tenure and efficient mechanisms for the registration, marketing, pricing, and transfer of parcels, together with the enforcement of contracts, are at a premium. Unambiguous title is also important: because urban land is so much more expensive than rural, financing is usually critical, for buyers and developers alike, and mortgage lenders like their investments to be secure. Indeed, the larger lenders prefer whole development packages to be standardized, because the risks are more predictable (Leinberger 2009). The result, as Evers and Korff (2000, 227) observe with respect to Bangkok, is that "the market emerges as the main mechanism for the structuration of [the city]." The same has been true for every urban site: they were the places where land first became a commodity, where the land market was first, and most closely, regulated and where it now frames the future.

But of course, historically and geographically, there must be a transition from rural land – unregulated, and perhaps not even viewed as a commodity – into the urban market. The places between city and country – the suburban areas and the peri-urban fringe that lies just beyond – are where that transition happens.

How Suburban Land Markets Differ

In some respects the land market in suburban areas is simply a watered-down version of that which exists in and around city centres (or, to use another analogy, when land becomes commodified, it is like planting seeds of a different kind from those used for agriculture). Land is less valuable, subject to fewer and less intense externalities, and less tightly regulated. There also is less infrastructure, notably in terms of public transit, although because of lower densities what there is may cost more to provide on a per capita basis (Blais 2010). In terms of market process as well as morphology, then, it can be thought of as "the city thinned out" (Douglass 1925, 3).

But when we consider the process by which the rural-urban transition happens, it becomes clear that the urban fringe presents some unique and significant features. The most striking and general one arises from the jump in value that occurs when rural land at the periphery is converted to urban use. In proportional terms, this is a far greater increase than in any other type of land-use change (Figure 1.2). Large profits are to be made by those who are able to buy cheap and sell dear, which can have a myriad of effects, depending on the context. The possibility gives landowners a strong incentive to develop their properties; where conversion is regulated, it encourages them to lobby or bribe local politicians to change the zoning; it attracts outside investors, whose deep pockets can up the ante. These days, India provides many well-documented examples, including Bangalore, Mumbai, and, as Raman shows in chapter 12, Chennai (Balakrishnan 2016; Gengaje 1992; Weinstein 2008). Local governments themselves are encouraged to claim some of the increase, whether through ad hoc bribery, or by claiming all or part of the "development gain" through the imposition of development charges and taxes on land transfers, or through appropriating land and then acting as developers in their own right (Dunkerley 1983, 23; Evans 2004b; Ingram and Hong 2012; Shoup 1983). The use of development charges, which are akin to the sort of land tax that has long been advocated – in recent years most effectively by the Lincoln Institute for Land Policy – has received relatively little attention from urban scholars (Haila 2016, 221; Ingram and Hong 2012). Historically, one of the most aggressive efforts to recoup development gain was made by the Labour government in post-war Britain but, unless governments actually own land, this strategy has proven very difficult to carry through (Cragoe and Readman 2010; Doebele 1983, 17). More recently, a version has been carried through very effectively by Singapore (Haila 2016). The most striking example of public ownership, however, is provided by China, where by 2013 municipalities obtained two-thirds of their revenue from land sales, mostly at the urban fringe (Cao 2015, 143; Wu 2015, 204–5). This issue is discussed by Wu and Li in chapter 6. In one form or another, then, the profits associated with suburban land conversion encourage the emergence of some type of "growth machine" that involves an exceptionally intimate relationship between private developers and local government (Logan and Molotch 1987).

Especially in parts of the global south, another distinctive element in the suburban land market concerns the manner in which the public infrastructure is provided. Today, in the north it is normal for hard infrastructure to be installed just before homes are built and occupied. In the south it has been common for the sequence to be reversed, as squatters settle, build, and then increase in number to the point where safe water and waste disposal must be provided (Baross 1990; Monstadt and Schramm 2013). Sometimes these services have been undertaken by the settlers themselves. Especially in recent years, governments have supported the process through programs of settlement upgrading (Wakely and Riley 2011). By the time upgrading happens the neighbourhoods in question may be urban in terms of density and even location. But in general the sequence of settlement-construction-servicing itself is a feature of the urban fringe.

Another suburban feature in certain parts of the world is the juxtaposition of private and customary forms of ownership. As cities push into rural territory, land conversion often entails a fundamental change in tenure, as well as of use. Sometimes traditional land rights are disregarded or otherwise dealt with "summarily" (Jenkins 2009, 9), and sometimes "mere occupation of land transforms it into a marketable commodity" (Baross and van der Linden 1990, 8), but usually the change is less sudden. It is complicated, being compromised by the emergence of hybrid practices and by the intermixture of various tenure forms, side by side (Evers and Korff 2000, 182–3). Razzaz (1994) provides a good example in the complicated interpenetration of western and Sharia tenure that occurred in the development of Yajouz, a suburb of Amman, Jordan. There, tribal claims had been overlain through appropriation by the British colonial state, but then urban expansion led to illegal squatting. The result was a combination of domination, complementarity, and conflict. In such situations, at least in the short run, political and cultural conflict is indeed predictable (Lombard and Rakodi 2016). In time the outcome, as described by Jenkins (2002, 216), may be a "general acceptance of modified traditional forms of land allocation in all but the inner 'cement cities.'"

In significant ways, then, the character of suburban land is different from that of its urban or rural counterparts. It is doubly transitional: it juxtaposes and combines very different tenure forms to an exceptional degree; and, straddling the zone that encompasses a fundamental change in land use, it offers the prospect of unusual speculative profits.

There are obviously some significant differences between suburban land markets in the global north and south, but broad generalizations about those ill-defined regions do not take us very far. In both world regions conditions vary. As Françoise Jarrige and her colleagues show in chapter 10, Montpellier, France, is not Denver, USA; as Andexlinger and his colleagues demonstrate in chapter 7, Austria is not the Netherlands.

The same goes for China if compared with India, or indeed anywhere. So the line between north and south is blurred and often arbitrary, a fact that is underlined by the way in which many nations have crossed it – notably Japan, then the Asian Tigers, and now, if falteringly, the BRICS (an association of Brazil, Russia, India, China, and South Africa). To begin to appreciate the subtler combinations that are currently apparent, it is necessary to consider what it is that makes the present period unique.

The Current Scene

Three broad trends are shaping suburban land markets around the world: urbanization, the globalization of capital and culture, and the steady commodification of land. A fourth change, a boom in land prices, is arguably more cyclical than secular in character, but that will become clear only in time.

The most immediate context for suburban land markets is urbanization, which involves not simply the growth of cities but an increase in the proportion of people living there. The two, of course, usually go together, and with worldwide population growth over the past century, together they have produced a rapid expansion of urban settlement (Angel 2012). Urbanization is a trend that had largely worked itself out in North America, northern Europe, Australasia, and Japan by 1945, although declining densities since then have meant that urban expansion has continued apace even here (Haase et al. 2011). In recent decades it has been more dramatic in other parts of the world (Angel 2012). In 1977 William Doebele (1977, 532) observed that "in absolute numbers, the flow of low-income persons to the urban areas of the less-developed countries since World War II constitutes one of the greatest migrations in human history." Since then the scope of change has widened while its pace has quickened. Globally, in the past quarter-century, there has been a more rapid rate of urbanization than in any previous period in history, the most striking changes being experienced by China (Champion and Hugo 2004; Cohen 2004).

The sheer rate of urbanization has mattered. Rapid growth has magnified some of the most distinctive features of the suburban market. It has increased the speed with which land values change, accelerating the dynamics of speculation and investment, putting owners, developers, and lenders under pressure. It has probably increased the proportion of land that is owned by investors (Brown, Phillips, and Roberts 1981, 137), and it has narrowed the window within which property has become commodified, a process that in many regions has hastened a reconciliation of private with customary forms of ownership. In all of these ways it has magnified the conflict and the challenges of governance (Doebele 1987a, 22). Rapid change has already created major problems for local governments. Even in affluent societies they have struggled in the past to regulate and service fast-growing suburban fringe areas. This was especially true where suburban government was weak or fragmented, so that rural jurisdictions had to adapt quickly to handle what soon became urban settlement. Toronto in the 1910s and Detroit in the 1920s and the 1940s present salutary examples of places where rapid adaptation came later (R. Harris 1996, 2004). Recently, the challenge has been many times greater in and around poorer cities, where governments have lacked the staff to enforce regulations or the resources to build infrastructure.

Since 1945, a common result of rapid urban growth has been extensive informal development. Some of this expansion has served local elites, who are able to disregard petty planning restrictions in their search for attractive or conveniently located sites, but most has reflected the needs of the poor. As Alan Gilbert recounts in chapter 9, squatter settlements flourished around Latin American cities when urbanization gathered momentum in the 1950s and 1960s (Klaufus and van Lindert 2012; Laquian 1964; Mangin 1967; Turner 1967). Subsequently, they became widespread in many parts of Asia and most recently in Africa, as these regions joined the trend (Berner 2012; Davis 2006; Neuwirth 2005). The other common type of informality, also pioneered on a large scale in Latin America and discussed by Gilbert, have been "pirate" settlements. Planned by entrepreneurs, they have typically involved the development of privately owned land in areas that government had not (yet) approved for settlement (Chabbi 1988; Doebele 1977). Informal settlement can take a great variety of forms, depending on local conditions. In many places it has produced shanty towns; in India shanty town residents are content to call them "slums," because they are then qualified for government assistance. In China, however, informal development appears as more substantial structures in villages that were rapidly encircled by urban growth, encapsulating a distinct mode of land tenure and encouraging densities to rise very high (Sun and Liu 2015). In whatever form, informal settlement has become one of the most distinctive elements of the suburban scene in many parts of the world.

Urban growth has reflected, and contributed to, a second trend: globalization (Spencer 2014). An extensive literature exists on this subject, which has many aspects: the internationalization of trade, capital, and to a lesser extent labour markets; and of culture, broadly defined to include consumer preferences and branding as well as tastes in food, clothing, and music. With increased globalization, English has become the lingua franca of business, trade, and to a lesser extent politics. In this context three interrelated elements have had an important effect on suburban land markets. The first concerns a standardization of the forms that legal, and indeed some informal, developments have taken. Conformity has been a feature of suburban development first in Britain and then in the United States since the late nineteenth century and was often criticized as such. Initially, it primarily affected the form of residential development – the long-idealized, low-density, middle-class suburb. However, since the 1970s and especially in the United States it has also helped to define commercial and office development, in large part because predictability is favoured by major mortgage lenders (Leinberger 2009). These models, particularly variants of the gated community, have been widely copied, from Brazil to India to South Africa and China (Glasze, Webster, and Frantz 2006; Herzog 2015).

This globalization of suburban forms has reflected two interrelated trends. To a considerable extent it has expressed a widespread popular desire to emulate, or to appropriate the cachet of, British and American models. This is apparent, for example, in the naming of these communities. In Beijing, for example, there are new projects called "American Gardens," "Vancouver Square," and "Paris Spring," while Shanghai can now boast "Harvard." Such practices are accepted, even welcomed, in many places, but Chinese pride has been piqued and the government is now trying to clamp down (*Economist* 2016). But they have also been the result of an internationalization of the people and organizations who are responsible, including architects, land developers, insurance companies, and mortgage lenders. Since the 1980s, these parties have become increasingly dominant as agents of development, as almost all postcolonial governments have taken a back seat, or at most a partnership role, in joint initiatives (see Levien 2013). Sometimes the results are good: in chapter 11 Margot Rubin and Richard Harris provide an example of a suburban development outside Johannesburg that has worked well to date. Too often, however, the state underwrites developers' profits, as Bhuvaneswari Raman shows for Chennai in chapter 12. Outsiders do not always carry the day. Searle (2014), for example, has documented some of the challenges for India in bringing together foreign investors with local developers. In that country, it seems, one of the sticking points has been disagreement about the valuation of land. But the general and long-term trend has been in the direction of an increase in the number of international linkages and probably of a greater concentration of economic power (Trivelli 1986). Major shocks to the system, notably the financial crisis of 2008, have dramatized but apparently not halted the trend.

Together, and for many decades, globalization and suburbanization have contributed to a third trend: the rapid transformation of property rights, in particular the growing commodification of land (Doebele 1987b, 116; Gough 1999; Jenkins 2009; Razzaz 1998). As argued previously, it is inevitable that suburban growth erodes customary tenure at the fringe. In recent decades the high stakes associated with rapid land conversion have attracted powerful, speculative investors who have brought pressure to bear on existing landowners and local governments. Resistance, such as that which some investors have faced in India, has been a rearguard action. International agencies, including the World Bank and the International Monetary Fund, have for many years promoted the establishment and consolidation of western-style markets in land and housing (R. Harris 2014). In recent decades, drawing, for example, on the ideas of Hernando de Soto (2000), they have advocated the "titling" of individual properties and by extension the regularization of informal settlements. The merits of these initiatives have been widely debated, but one effect is clear: the reach and hold of private property markets has steadily grown.

It appears that, associated with these trends in many parts of the world, there has been a steady increase in the relative price of all types of urban land (*Economist* 2015; see Doebele 1987b, 116; Dunkerley 1983, 10). No firmer statement is possible because comparable international data are not available, while land markets continue to be volatile and local. Evidence presented recently by Piketty indicates that, while the value of agricultural land as a proportion of GDP has been declining in the global north since the eighteenth century, the value of urban residential property has soared, especially since the 1960s (*Economist* 2015). Fragmentary evidence indicates that similar trends have affected other regions. Latin America was an early example, where Trivelli (1986) has shown that rates of increase in property values in the post-war era were greatest in suburban areas, a fact that he attributes to widespread speculation. More recent fluctuations and secular increases on a global scale have been tied to international movements of capital and to the perceived profitability and security of investments in land. Indeed, a strong case can be made that speculation in land, especially that located in suburban areas, is not only an expression of recent trends in global finance but a significant factor in its evolution. The example of the subprime crisis in the United States is the most dramatic of these trends to North Americans, but the extraordinary property boom that has reshaped Chinese cities since the 1980s is surely of equal importance, as Fulong Wu and Zhigang Li's chapter 6 suggests. Unravelling these connections is beyond our ability and the scope of this introduction, but it will be vital to our future understanding of the development of land markets in suburban areas.

As our survey indicates, the suburban land question embraces a very wide range of issues, cultural, economic, and political. They might be classified in a variety of ways, but we have chosen to order the papers in this collection approximately according to their respective emphases on process, form, or planning. In so doing we have been guided by our understanding of the transitional character of suburban territory, shaped as it is by social processes of many kinds, resisted by existing patterns of land use and ownership, and in both respects guided to a greater or lesser extent by collective efforts for land planning.

In this introduction, where we have tried to provide a systematic treatment of the subject, we have indicated the particular aspect that each contributor speaks to most directly. But in truth, many chapters straddle two or more aspects, which reflects the fact that there is a continual interplay between process and form, and that in the modern era neither can be understood without reference to their regulatory context. For that reason we have not tried to define distinct sections. Instead, we have arranged the chapters on a rough continuum, as the emphasis shifts from a concern with process and form towards a discussion of planning and governance. The one mode of chapter organization that we resisted on principle was by region. As we have argued, although the suburban land question takes on a different character in each city, nation, and world region, it also displays some fundamental similarities that transcend borders. It is, above all, to those commonalities that we wish to draw attention.

REFERENCES

Ajl, M. 2014. "The Hypertrophic City versus the Planet of Fields." In *Implosions/Explosions: Towards a Study of Planetary Urbanization*, ed. N. Brenner, 533–50. Berlin: Jovis.

AlShehabi, O.H., and S. Suroor. 2016. "Unpacking 'Accumulation by Dispossession,' 'Fictitious Commodification,' and 'Fictitious Capital Formation': Tracing the Dynamics of Bahrain's Land Reclamation. *Antipode* 48 (4): 835–56. https://doi.org/10.1111/anti.12222.

Andersson, D.E., and S. Moroni, eds. 2014. *Cities and Private Planning. Property Rights, Entrepreneurship and Transaction Costs*. Cheltenham, UK: Edward Elgar.

Angel, S. 2012. *Planet of Cities*. Cambridge, MA: Lincoln Institute of Land Policy.

Angel, S., J. Parent, D.L. Civco, and A.M. Blei. 2012. *Atlas of Urban Expansion*. Cambridge, MA: Lincoln Institute for Land Policy.

Angelo, H. 2017. "From the City Lens toward Urbanisation as a Way of Seeing: Country/City Binaries on an Urbanising Planet." *Urban Studies* 54 (1): 158–78. https://doi.org/10.1177/0042098016629312.

Arboleda, M. 2015. "Spaces of Extraction, Metropolitan Explosions: Planetary Urbanization and the Commodity Boom in Latin America." *International Journal of Urban and Regional Research*. https://doi.org/10.1111/1468-2427.12290.

Balakrishnan, S. 2016. "Periurban Land Markets in the Bangalore Region." In *Slums: How Informal Real Estate Markets Work*, ed. E.L. Birch, S. Chattaraj, and S.M. Wachter, 107–20. Philadelphia: University of Pennsylvania Press. https://doi.org/10.9783/9780812292572-008.

Baross, P. 1990. "Sequencing Land Development. The Price Implications of Legal and Illegal Settlement Growth." In *The Transformation of Land Supply Systems in Third World Cities*, ed. P. Baross and J. van der Linden, 57–87. Aldershot, UK: Avebury.

Baross, P., and J. van der Linden. 1990. "Introduction." In *The Transformation of Land Supply Systems in Third World Cities*, ed. P. Baross and J. van der Linden, 1–16. Aldershot, UK: Avebury. https://doi.org/10.1016/0077-7579(90)90066-P.

Berner, E. 2012. "Informal Housing: Asia." In *International Encyclopedia of Housing and Home*, ed. S. Smith, 56–77. Amsterdam: Elsevier. https://doi.org/10.1016/B978-0-08-047163-1.00700-1.

Bertaud, A. 2010. "Land Markets, Government Interventions, and Housing Affordability." Working Paper 18. Wolfensohn Center for Development, Washington, DC. http://alainbertaud.com/wp-content/uploads/2013/08/Bertaud-_land_Markets_FINAL_1a.pdf.

Bertaud, A., and B. Renaud. 1997. "Socialist Cities without Land Markets." *Journal of Urban Economics* 41 (1): 137–51. https://doi.org/10.1006/juec.1996.1097.

Blais, P. 2010. *Perverse Cities: Hidden Subsidies, Wonky Policies, and Urban Sprawl.* Vancouver: UBC Press.

Block, F., and P. Evans. 2005. "The State and the Economy." In *The Handbook of Economic Sociology*, ed. N.J. Smelser and R. Swedberg, 505–26. Princeton, NJ: Princeton University Press.

Brenner, N. 2013. "Theses on Urbanization." *Public Culture* 25 (1 69): 85–114. https://doi.org/10.1215/08992363-1890477.

Brenner, N. 2014. "Introduction: Urban Theory without an Outside" in *Implosions/Explosions: Towards a Study of Planetary Urbanization*, ed. N. Brenner, 14–30. Berlin: Jovis.

Brenner, N., and C. Schmid. 2014. "The 'Urban Age' in Question." *International Journal of Urban and Regional Research* 38 (3): 731–55. https://doi.org/10.1111/1468-2427.12115.

Brenner, N., and C. Schmid. 2015. "Towards a New Epistemology of the Urban?" *City* 19 (2–3): 151–82. https://doi.org/10.1080/13604813.2015.1014712.

Brown, H.J., and N.A. Roberts. 1978. "Land Owners at the Urban Fringe," Urban Planning, Policy Analysis, and Administration. Discussion Paper 78–10, Department of City and Regional Planning, Harvard University.

Brown, H.J., R.S. Phillips, and N.A. Roberts. 1981. "Land Markets at the Urban Fringe: New Insights for Policy Makers." *Journal of the American Planning Association* 47 (2): 131–44. https://doi.org/10.1080/01944368108977098.

Bruce, J.W. 1988. "A Perspective on Indigenous Land Tenure Systems and Land Concentration." In *Land and Society in Contemporary Africa*, ed. R.E. Downs and S.P. Reyna, 23–52. Hanover, NH: University Press of New England.

Cao, J. 2015. *Chinese Real Estate Market: Development, Regulation and Investment.* London: Routledge.

Capaldi, C.A., Dopko, R.L., and Zelenski, J.M. 2014. "The Relationship between Nature Connectedness and Happiness: A Meta-Analysis." *Frontiers in Psychology* 5 (8 September): 976. https://doi.org/10.3389/fpsyg.2014.00976.

Chabbi, M. 1988. "The Pirate Subdeveloper. A New Form of Land Developer in Tunis." *International Journal of Urban and Regional Research* 12 (1): 8–21. https://doi.org/10.1111/j.1468-2427.1988.tb00071.x.

Champion, A.G., and G. Hugo, eds. 2004. *New Forms of Urbanisation: Beyond the Rural-Urban Dichotomy.* Aldershot, UK: Ashgate.

Christaller, W. 1933. *Zentrale Orte in Süddeutschland.* Jena: Gustav Fischer.

Christophers, B. 2017. "The State and Financialization of Public Land in the United Kingdom." *Antipode* 49 (1): 62–85. https://doi.org/10.1111/anti.12267.

Clapson, M. 2003. *Suburban Century: Social Change and Urban Growth in England and the USA*. Oxford: Berg.

Cohen, B. 2004. "Urban Growth in Developing Countries. A Review of Current Trends and a Caution Regarding Existing Forecasts." *World Development* 32 (1): 23–51. https://doi.org/10.1016/j.worlddev.2003.04.008.

Cragoe, M., and P. Readman, eds. 2010. *The Land Question in Britain, 1750–1950*. New York: Palgrave Macmillan. https://doi.org/10.1057/9780230248472.

Cribbet, J. 1963. "Condominium: Homeownership for Megalopolis?" *Michigan Law Review* 61 (7): 1207–44. https://doi.org/10.2307/1286564.

Crook, T., J. Henneberry, and C. Whitehead. 2016. *Planning Gain. Providing Infrastructure and Affordable Housing*. Chichester, UK: Wiley.

Davis, M. 2006. *Planet of Slums*. London: Verso.

de Soto, H. 2000. *The Mystery of Capital*. New York: Basic Books.

de Sousa Santos, B. 1977. "The Law of the Oppressed: The Construction and Reproduction of Legality in Pasagarda." *Law & Society Review* 12 (1): 5–126. https://doi.org/10.2307/3053321.

Doebele, W.A. 1977. "The Private Market and Low Income Urbanisation. The 'Pirate' Subdivisions of Bogotá." *American Journal of Comparative Law* 25 (3): 531–64. https://doi.org/10.2307/839692.

Doebele, W.A. 1983. "Concepts of Urban Land Tenure." In *Urban Land Policy. Issues and Opportunities*, ed. H.B. Dunkerley, 63–107. New York: Oxford University Press.

Doebele, W.A. 1987a. "The Evolution of Concepts of Urban Land Tenure in Developing Countries." *Habitat International* 11 (1): 7–22. https://doi.org/10.1016/0197-3975(87)90030-0.

Doebele, W.A. 1987b. "Land Policy." In *Shelter, Settlement and Development*, ed. L. Rodwin, 110–32. London: Allen & Unwin.

Douglas, I. 2006. "Peri-urban Ecosystems and Societies: Transitional Zones and Contrasting Values." In *The Peri-urban Interface*, ed. D. McGregor et al., 18–29. London: Earthscan.

Douglass, H. 1925. *The Suburban Trend*. New York: Century.

Dunkerley, H.B., ed. 1983. *Urban Land Policy: Issues and Opportunities*. New York: Oxford University Press.

Durand-Lasserve, A., M. Durand-Lasserve, and H. Selod. 2015. *Land Delivery Systems in West African Cities: The Example of Bamako, Mali*. Washington, DC: World Bank; https://openknowledge.worldbank.org/bitstream/handle/10986/21613/9781464804335.pdf?sequence=1. Accessed 30 June 2016. https://doi.org/10.1596/978-1-4648-0433-5.

Economist, The. 2015. "The Paradox of Soil." 14 April. Accessed 30 June 2016. https://www.economist.com/news/

briefing/21647622-land-centre-pre-industrial-economy-has-returned-constraint-growth.
Economist, The. 2016. "Exterminate the Foreign Names." 7 May. Accessed 30 June 2016. https://www.economist.com/news/china/21698293-xi-jinpings-latest-purge-exterminate-foreign-names.
Ekers, M., P. Hamel, and R. Keil. 2012. "Governing Suburbia: Modalities and Mechanisms of Suburban Governance." *Regional Studies* 46 (3): 405–22. https://doi.org/10.1080/00343404.2012.658036.
Evans, A.W. 2004a. *Economics, Real Estate and the Supply of Land.* Oxford: Blackwell.
Evans, A.W. 2004b. *Economics and Land Use Planning*. Oxford: Blackwell. https://doi.org/10.1002/9780470690895.
Evers, H.-D., and R. Korff. 2000. *Southeast Asian Urbanism. The Meaning and Power of Social Space*. New York: St Martin's Press.
Fallows, J. 2016. "How America Is Putting Itself Back Together." *The Atlantic*, March. Accessed 30 June 2016. https://www.theatlantic.com/magazine/archive/2016/03/how-america-is-putting-itself-back-together/426882/.
Fitzgibbon, D. 2011. "Assembling the Property Market in Imperial Britain, c. 1750-1925." PhD dissertation, University of California, Berkeley.
Friedmann, J. 2014. "Becoming Urban: On Whose Terms?" In *Implosions/Explosions: Towards a Study of Planetary Urbanization*, ed. N. Brenner, 551–65. Berlin: Jovis.
Fureseth, O.J., and M.B. Lapping. 1999. *Contested Countryside. The Rural-Urban Fringe in North America*, Aldershot, UK: Ashgate.
Gandy, M. 2012. "Where Does the City End?" *Architecture Design* 82 (1): 128–32. https://doi.org/10.1002/ad.1363.
Gans, H. 1969. "Urbanism and Suburbanism as Ways of Life." In Gans, H., *Essays on Urban Problems and Solutions*. New York: Basic Books.
Garreau, J. 1991. *Edge City: Life on the New Frontier*. New York: Doubleday.
Gengaje, R.K. 1992. "Administration of Farmland Transfer in Urban Fringes: Lessons from Maharashstra, India." *Land Use Policy* 9 (4): 272–86. https://doi.org/10.1016/0264-8377(92)90004-G.
Ginsburg, N. 1991. "Extended Metropolitan Regions in Asia: A New Spatial Paradigm." In *The Extended Metropolis. Settlement Transition in Asia*, ed. N. Ginsburg, B. Koppell, and T.M. McGee, 27–46. Honolulu: University of Hawaii Press.
Glasze, G., C. Webster, and K. Frantz, eds. 2006. *Private Cities: Global and Local Perspectives*. New York: Routledge.
Goddard, J. 2012. *Being American on the Edge: Penurbia and the Metropolitan Mind, 1945–2010*. New York: Palgrave.

Goonewardena, Kanishka. 2014. "The Country and the City in the Urban Revolution." *Implosions/Explosions: Towards a Study of Planetary Urbanization*, ed. Neil Brenner, 218–31. Berlin: Jovis.

Gough, K.V. 1999. "The Changing Role and Nature of Urban Governance in Peri-Urban Accra, Ghana." *Third World Planning Review* 21 (4): 393–410. https://doi.org/10.3828/twpr.21.4.465l0m84v32q8103.

"Grand River Land Dispute." 2016. Accessed 9 June 2016. https://en.wikipedia.org/wiki/Grand_River_land_dispute.

Haase, A., A. Steinführer, S. Kabisch, K. Grossmann, and R. Hall, eds. 2011. *Residential Change and Demographic Challenge: The Inner City of East Central Europe in the 21st Century*. Aldershot, UK: Ashgate.

Haila, A. 2016. *Urban Land Rent: Singapore as a Property State*. New York: Wiley.

Hall, P., and K. Pain. 2006. *The Polycentric Metropolis: Learning from Mega-City Regions in Europe*. London: Earthscan.

Harris, C. 1992. "The Lower Mainland, 1820–1881." In *Vancouver and Its Region*, ed. G. Wynn and T. Opke, 38–68. Vancouver: UBC Press.

Harris, D. 2011. "Condominium and the City: The Rise of Property in Vancouver." *Law & Social Inquiry* 36 (3): 694–726. https://doi.org/10.1111/j.1747-4469.2011.01247.x.

Harris, R. 1996. *Unplanned Suburbs: Toronto's American Tragedy, 1900–1950*. Baltimore: Johns Hopkins University Press.

Harris, R. 2004. *Creeping Conformity: How Canada Became Suburban, 1900–1960*. Toronto: University of Toronto Press.

Harris, R. 2014. "Urban Land Markets: A Southern Exposure." In *The Routledge Handbook on Cities of the Global South*, ed. S. Parnell and S. Oldfield, 109–21. London: Routledge.

Harris, R. 2015. "Using Toronto to Explore Three Suburban Stereotypes, and Vice Versa." *Environment and Planning A* 47 (1): 30–49.

Harris, R. 2017. "Modes of Informal Urban Development: A Global Phenomenon." *Journal of Planning Literature*. https://doi.org/10.1177/0885412217737340

Harris, R., and R. Lewis. 2001. "The Geography of North American Cities and Suburbs, 1900–1950: A New Synthesis." *Journal of Urban History* 27 (3): 262–92. https://doi.org/10.1177/009614420102700302.

Harris, R., U. Lehrer, and R. Bloch. 2013. "The Suburban Land Question." Discussion paper, Global Suburbanisms Project Workshop on "Land." Montpellier, France, 21-23 October 2012. Rev. MS.

Herzog, L.A. 2015. *Global Suburbs. Urban Sprawl from the Rio Grande to Rio de Janeiro*. London: Routledge.

Howard, E. 1898. *To-Morrow: A Peaceful Path to Real Reform*. 1st ed. London: Swan Sonnenschein. Retitled *Garden Cities of To-Morrow*. 1902. 2nd ed. London: Swan Sonnenschein. Many subsequent eds to 2007.

Hoyt, H. 1933. *One Hundred Years of Land Values in Chicago*. Chicago: University of Chicago Press.

Ingram, G.K., and Y.-H. Hong, eds. 2012. *Value Capture and Land Policies*. Cambridge, MA: Lincoln Institute of Land Policy.

Jackson, K. 1985. *Crabgrass Frontier: The Suburbanization of the United States*. New York: Oxford University Press.

Jacobs, H.M., and K. Paulsen. 2009. "Property Rights. The Neglected Theme of 20th-Century American Planning." *Journal of the American Planning Association* 75 (2): 134–43. https://doi.org/10.1080/01944360802619721.

Jenkins, P. 2002. "Beyond the Formal/Informal Dichotomy: Access to Land in Maputo, Mozambique." In *Reconsidering Informality. Perspectives from Urban Africa*, ed. K.T. Hansen and M. Vaa, 210–26. Oslo: Nordiska Afrikainstitutet.

Jenkins, P. 2009. "African Cities: Competing Claims on Urban Land." In *African Cities: Competing Claims over Urban Spaces*, ed. F. Locatelli and P. Nugent, 81–108. Leiden: Brill. https://doi.org/10.1163/ej.9789004162648.i-308.29.

Johnston, R.J. 2000. "Von Thünen Model." In *The Dictionary of Human Geography*, ed. R.J. Johnston, D. Gregory, G. Pratt, and M. Watts, 895–6. Oxford: Blackwell.

Kaika, M., and L. Ruggiero. 2016. "Land Financialization as a 'Lived' Process: The Transformation of Milan's Bicocca by Pirelli." *European Urban and Regional Studies* 23 (1): 3–22. https://doi.org/10.1177/0969776413484166.

Karr, R.D. 2015. "Suburban Land Development in Antebellum Boston." *Journal of Urban History* 41 (5): 862–80. https://doi.org/10.1177/0096144214566955.

Keil, R. 2017. *Suburban Planet: Making the World Urban from the Outside In*. Cambridge: Polity.

Keil, R. 2018. "Constructing Global Suburbia, One Critical Theory at a Time." In *Doing Global Urban Research*, ed. J. Harrison and M. Hoyler, 169–81. Thousand Oaks, CA: Sage.

Keil, R., ed. 2013. *Suburban Constellations: Governance, Land and Infrastructure in the 21st Century*. Berlin: Jovis.

Keil, R., and J. Graham. 1998. "Constructing Urban Environments after Fordism," in *Remaking Reality: Nature at the Millennium*," ed. N. Castree and B. Braun, 100–25. London and New York: Routledge.

Kimani, S.M. 1972. "Spatial Structure of Land Values in Nairobi." *Tijdschrift voor Economische en Sociale Geografie* 63 (2): 105–14. https://doi.org/10.1111/j.1467-9663.1972.tb01173.x.

Klaufus, C., and P. van Lindert. 2012. "Informal Housing: Latin America." In *International Encyclopedia of Housing and Home*, ed. S. Smith, 70–7. Amsterdam: Elsevier. https://doi.org/10.1016/B978-0-08-047163-1.00482-3.

Knox, P.L. 1991. "The Restless Urban Landscape: Economic and Sociocultural Change and the Transformation of Metropolitan Washington, DC." *Annals of the Association of American Geographers* 81 (2): 181–209. https://doi.org/10.1111/j.1467-8306.1991.tb01686.x.

Krueckeberg, D.A. 1999. "Private Property in Africa. Creation Stories of Economy, State and Culture." *Journal of Planning Education and Research* 19 (2): 176–82. https://doi.org/10.1177/0739456X9901900207.

Laquian, A. 1964. "Isla de Kokomo. Politics among Urban Slum Dwellers." *Philippine Journal of Public Administration* 8 (2): 112–22.

Lefebvre, H. 2003. *The Urban Revolution*. Trans. Robert Bononno. Minneapolis: University of Minnesota Press.

Le Goix, R. 2016. "L'immobilier résidentiel suburbain en régime financiarisé de production dans la région de Los Angeles." *Revue d'économie régionale et urbaine* (ed. Armand Colin) 1: 101–30.

Lehrer, U. 1994. "Images of the Periphery: The Architecture of FlexSpace in Switzerland." *Environment and Planning. D, Society & Space* 12 (2): 187–205. https://doi.org/10.1068/d120187.

Lehrer, U. 2012. ""If you lived here …': Lifestyle, Marketing and Development of Condominiums in Toronto." *Scapegoat* 3: 23.

Lehrer, U. 2013. "FlexSpace – Suburban Forms" in *Suburban Constellations: Governance, Land and Infrastructure in the 21st Century*, ed. R. Kiel, 56–61. Berlin: Jovis.

Lehrer, U. 2016. "Room for the Good Society? Public Space, Amenities and the Condominium." In *Insurgencies and Revolutions: Reflections on John Friedmann's Contributions to Planning Theory and Practice*, ed. Haripriya Rangan, Mee Kam Ng, Libby Porter, and Jacquelyn Chase. RTPI Library Series. Boston: Routledge.

Lehrer, U., and T. Wieditz. 2009. "Condominium Development and Gentrification: The Relationship between Policies, Building Activities and Socio-economic Development in Toronto." *Canadian Journal of Urban Research* 18 (1): 82–103.

Leinberger, C. 2009. *The Option of Urbanism. Investing in a New American Dream*. Washington, DC: Island Press.

Levien, M. 2013. "Regimes of Dispossession: From Steel Towns to Special Economic Zones." *Development and Change* 44 (2): 381–407. https://doi.org/10.1111/dech.12012.

Logan, J., and H. Molotch. 1987. *Urban Fortunes: The Political Economy of Place.* Toronto: University of Toronto Press.

Lombard, M., and C. Rakodi. 2016. "Urban Land Conflict in the Global South: Towards an Analytical Framework." *Urban Studies* 53 (13): 2683–99. https://doi.org/10.1177/0042098016659616.

Macdonald, S., and R. Keil. 2012. "The Ontario Greenbelt: Shifting the Scales of the Sustainability Fix." *Professional Geographer* 64 (1): 125–45. https://doi.org/10.1080/00330124.2011.586874.

Mangin, W. 1967. "Latin American Squatter Settlements: A Problem and a Solution." *Latin American Research Review* 2 (3): 65–98.

McKenzie, E. 1994. *Privatopia: Homeowner Associations and the Rise of Residential Private Government.* New Haven: Yale University Press.

McKenzie, E. 2011. *Beyond Privatopia: Rethinking Residential Private Government.* Washington, DC: Urban Institute.

McManus, R., and P. Ethington. 2007. "Suburbs in Transition: New Approaches to Suburban History." *Urban History* 34 (2): 317–37. https://doi.org/10.1017/S096392680700466X.

Mohan, R. 1994. *Understanding the Developing Metropolis. Lessons from the City Study of Bogota and Cali, Colombia.* Oxford: Oxford University Press.

Monte-Mór, R.L. 2005. "What Is the Urban in the Contemporary World?" *Cad. Saúde Pública, Rio De Janeiro* 21 (3): 942–8. https://doi.org/10.1590/S0102-311X2005000300030.

Monstadt, J., and S. Schramm. 2013. "Beyond the Networked City? Suburban Constellations in Water and Sanitation Systems." In *Suburban Constellations. Governance, Land and Infrastructure in the 21st. Century*, ed. R. Keil, 85–94. Berlin: Jovis.

Neuwirth, R. 2005. *Shadow Cities: A Billion Squatters. A New World.* New York: Routledge.

Nissim, R. 2008. *Land Administration and Practice in Hong Kong.* Hong Kong: Hong Kong University Press.

Porter, L., and J. Barry. 2016. *Planning for Coexistence? Recognizing Indigenous Rights through Land-use Planning in Canada and Australia.* London: Routledge.

Portes, A., and W. Haller. 2005. "The Informal Economy." In *The Handbook of Economic Sociology*, ed. N.J. Smelser and R. Swedberg, 403–25. Princeton, NJ: Princeton University Press.

Razzaz, O. 1994. "Contestation and Mutual Adjustment. The Process of Controlling Land in Yajouz, Jordan." *Law & Society Review* 28 (2): 1–39.

Razzaz, O. 1998. "Land Disputes in the Absence of Ownership Rights: Insights from Jordan." In *Illegal Cities. Law and Urban Change in Developing Countries*, ed. E. Fernandes and A. Varley, 69–88. London: Zed.

Robinson, J. 2006. *Ordinary Cities: Between Modernity and Development*. Abingdon, UK: Routledge.

Robinson, J. 2011. "Cities in a World of Cities: The Comparative Gesture." *International Journal of Urban and Regional Research* 35 (1): 1–23. https://doi.org/10.1111/j.1468-2427.2010.00982.x.

Roy, A. 2005. "Urban Informality: Towards an Epistemology of Planning." *Journal of the American Planning Association* 71 (2): 147–58. https://doi.org/10.1080/01944360508976689.

Roy, A. 2009. "Why India Cannot Plan Its Cities: Informality, Insurgence and the Idiom of Urbanization." *Planning Theory* 8 (1): 76–87. https://doi.org/10.1177/1473095208099299.

Roy, A. 2011. "Urbanisms, Worlding Practices and the Theory of Planning." *Planning Theory* 10 (1): 6–15. https://doi.org/10.1177/1473095210386065.

Roy, A. 2015. *The Land Question*. LSE Cities. https://lsecities.net/media/objects/articles/the-land-question/en-gb/. Accessed 30 June 2016.

Sandberg, L.A., G. Wekerle, and L. Gilbert. 2013. *The Oak Ridges Moraine Battles: Development, Sprawl, and Nature Conservation in the Toronto Region*. Toronto: University of Toronto Press.

Savage, L., and M. Lapping. 2003. "Sprawl and Its Discontents. The Rural Dimension." In *Suburban Sprawl: Culture, Theory, Politics*, ed. M. Lindstrom and H. Bartling, 5–17. Lanham, MD: Rowman & Littlefield.

Schmid, C. 2005. *Stadt, Raum und Gesellschaft: Henri Lefebvre und die Theorie der Produktion des Raumes*. Stuttgart: Steiner.

Schmid, C. 2006. "Theory." In *P. de Meuron and C. Schmid, Switzerland: An Urban Portrait*, Vol. 1, ed. R. Diener, J. Herzog, and M. Meili, 163–221. Basel: ETH Studio Basel, Birkhäuser.

Schmid, C. 2012. "Pattern and Pathways of Global Urbanization: Towards Comparative Analysis." In *Globalization of Urbanity*, ed. J. Acebillo, 51–77. Barcelona: Università della Svizzera Italiana and Actar.

Schönig, B., ed. 2015. *Variations of Suburbanism: Approaching a Global Phenomenon*. Stuttgart: Steiner.

Searle, L.G. 2014. "Conflict and Commensuration: Contested Market Making in India's Private Real Estate Development Sector." *International Journal of Urban and Regional Research* 38 (1): 60–78. https://doi.org/10.1111/1468-2427.12042.

Sevilla-Buitrago, A. 2014. "*Urbs in Rure*: Historical Enclosure and the Extended Urbanization of the Countryside." In *Implosions/Explosions: Towards a Study of Planetary Urbanization*, ed. N. Brenner, 236–59. Berlin: Jovis.

Shen, J., and F. Wu. 2017. "The Suburb as a Space of Capital Accumulation: The Development of New Towns in Shanghai, China." *Antipode* 49 (3): 761–80. https://doi.org/10.1111/anti.12302.

Shorto, R. 2013. *Amsterdam: A History of the World's Most Liberal City*. New York: Vintage.
Shoup, D.C. 1983. "Intervention through Property Taxation and Public Ownership." In *Urban Land Policy: Issues and Opportunities*, ed. H.B. Dunkerley, 132–52. New York: Oxford University Press.
Sieverts, T. 2002. *Cities without Cities: Between Place and World, Space and Time, Town and Country*. London: Routledge.
Simon, D. 2008. "Urban Environments. Issues on the Peri-Urban Fringe." *Annual Review of Environment and Resources* 33 (1): 167–85. https://doi.org/10.1146/annurev.environ.33.021407.093240.
Simpson, J.R. 1976. *Land Law and Registration*. Cambridge: Cambridge University Press.
Soja, E. 1992. "Inside Exopolis: Scenes from Orange Country" in *Variations on a Theme Park*, ed. M. Sorkin, 94–122. New York: Noonday Press.
Sparkes, P. 2003. *A New Land Law*. 2nd ed. Oxford: Hart.
Spencer, J.H. 2014. *Globalisation and Urbanisation: The Global Urban Ecosystem*. Manham, MD: Rowman & Littlefield.
Springett, J. 1982. "Landowners and Urban Development: The Ramsden Estate and Nineteenth-Century Huddersfield." *Journal of Historical Geography* 8 (2): 129–44. https://doi.org/10.1016/0305-7488(82)90003-2.
Sun, L. and Liu, Z. 2015. "Illegal But Rational: Why Small Property Rights Housing Is Big in China." *Land Lines* (July): 14–19.
Tamanha, B.Z. 2000. "A Non-Essentialist Version of Legal Pluralism." *Journal of Law and Society* 27 (2): 296–321. https://doi.org/10.1111/1467-6478.00155.
Tomlinson, K. 2016. "On B.C.'s Farmland, Mega-Mansions and Speculators Reap the Rewards of Lucrative Tax Breaks." *Globe and Mail*, 23 November. https://www.theglobeandmail.com/news/investigations/farmland-and-real-estate-in-british-columbia/article32923810/
Trivelli, P. 1986. "Access to Land by the Urban Poor." *Land Use Policy* 3 (2): 101–21. https://doi.org/10.1016/0264-8377(86)90048-7.
Turner, J.F.C. 1967. "Barriers and Channels for Housing Development in Modernizing Countries." *Journal of the American Institute of Planners* 33 (3): 167–81. https://doi.org/10.1080/01944366708977912.
Wakely, P., and E. Riley. 2011. *Cities without Slums. The Case for Incremental Housing*. Washington, DC: Cities Alliance.
Walks, A. 2013. "Suburbanism as a Way of Life, Slight Return." *Urban Studies* 50 (8): 1471–88. https://doi.org/10.1177/0042098012462610.
Walks, R.A. 2008. "Urban Form, Everyday Life, and Ideology: Support for Privatization in Three Toronto Neighborhoods." *Environment & Planning A* 40 (2): 258–82. https://doi.org/10.1068/a3948.

Weaver, J. 2003. *The Great Land Rush and the Making of the Modern World, 1650–1900*. Montreal and Kingston: McGill-Queen's University Press.
Weightman, B.A. 1972. "The Musqueam Reserve: A Case Study of the Indian Social Milieu in an Urban Environment." PhD. dissertation, University of Washington.
Weinstein, L. 2008. "Mumbai's Development Mafias: Globalisation, Organized Crime and Land Development." *International Journal of Urban and Regional Research* 32 (1): 22–39. https://doi.org/10.1111/j.1468-2427.2008.00766.x.
Wu, F. 2015. *Planning for Growth: Urban and Regional Planning in China*. Abingdon, UK: Routledge.
Young, D., P.B. Wood, and R. Keil, eds. 2011. *In-Between Infrastructure. Urban Connectivity in an Age of Vulnerability*. Kelowna, BC: Praxis Press.
Zheng, S., and M.E. Kahn. 2008. "Land and Residential Property Markets in a Booming Economy: New Evidence from Beijing." *Journal of Urban Economics* 63 (2): 743–57.

chapter two

Alternative Peripheries: Socialist Mass Housing Compared with Modern Suburbia

SONIA A. HIRT

In the aftermath of World War II vast chunks of land around large cities across the Soviet Union and the other Eastern-bloc countries were taken over by districts of large collectivist residential structures, constructed with industrialized building methods along the principles of modernist architecture and urban design. Known as *mikrorayoni* (микрорайони) (literally, "micro-regions" in Russian) in the USSR or "residential complexes" in some of the other Eastern-bloc countries, these mass-housing districts could well be described as the signature contribution of state socialism to the city. Their scale and ambition exceeded those of their look-alike contemporaries in Western cities, whether the French *grands ensembles*, the Swedish residential areas under the "Million Programme," or the American Pruitt-Igoe-type urban renewal projects. In the Russian capital, Moscow, over 80 per cent of the housing stock was built roughly between 1960 and 2000, the overwhelming majority of it in standardized mass-housing districts surrounding the historic city core (Krasheninnokov 2003). Among the capitals of the other Eastern-bloc countries only Bucharest, Romania, exceeded the 80 per cent mark. Elsewhere the percentage of socialist-era mass-housing out of the total dwelling stock was a bit lower, likely because the state-led push to relocate populations into large cities was weaker; for instance, about 55 per cent of the dwelling stock in Warsaw and 60 per cent in Sofia comprised standardized units (Hirt and Stanilov 2009). Still, this share far surpassed the percentage of similar housing in the large cities of capitalist countries – certainly in the United States, where the majority of the housing stock built during the mid-twentieth century took the form of low-density suburban single-family homes.

The physical differences between the "typical" Soviet apartment building and the "typical" Levittown-like American single-family home are obvious: the two can easily be viewed as architectural antipodes. Not surprisingly, several authors, the present one included, have doubted that the areas they dominated can ever be put under the same label, for example, "suburbs," if we understand this term to imply a certain physical form (Gornostayeva 1991; Tammaru 2001; Hirt 2007b). Still, the apparent antipodes have some basic things in common. Both are the paradigmatic housing districts that surrounded the mid-twentieth-century city and served as its alternative. Both are primary examples of decentred, industrial-age, urban expansion, and both reflect certain ideological imperatives that were shared across the Iron Curtain countries: belief in the power of all-transformative, comprehensive planning; physical determinism; a deep admiration for scientific management and new building technologies; and an aversion to the existing city centre perceived as an overcrowded and old-fashioned place, unbefitting to modern society.

This chapter discusses the broad processes and patterns of residential development in the periphery of socialist, primarily Soviet, urban cores during the mid-twentieth century with reference to the simultaneous processes and patterns of residential development in capitalist, primarily US metropolitan, settings. The discussion covers the decades from roughly 1930, by which time the communist state was firmly entrenched, to 1980, which marked the climax of the Soviet system of governance. The contrasts between the housing and neighbourhood types in these two contexts have been discussed much more than the similarities. Yet are there similarities? If we can demonstrate meaningful kinship between the most common forms of housing growth in the outskirts of cities that were produced by the two major rival socio-economic orders of the twentieth century – Soviet socialism and American capitalism – then we can add fuel to the thesis that twentieth-century suburban development can be viewed as a single, worldwide event. I organize my reflections along three aspects of development pertaining to the periphery of mid-twentieth-century cities: places, processes, and philosophies. My proposal is that, although along all three there are sharp contrasts between the socialist and the capitalist cases, some fundamental parallels exist. The latter stem from the fact that both the Soviet and the American versions of mid-twentieth-century housing development at the urban edge were an essentially modernist endeavour aimed at the transformation of urban form and lifestyles.

Places

A logical place to start a discussion on whether any form of urbanization, including Soviet-style mass housing, would qualify as suburban is to consider the questions of "when" and "where." In an excellent review of the global variety of suburban forms and definitions, Richard Harris (2010) points out that there are very few aspects of suburbia that are meaningful worldwide. Two of them concern time and place: suburbs are typically deemed as those places that are relatively new in a given metropolitan area and those places that are, at the moment of their "newness," located at the edge of the city. However, as growing cities tend to absorb their peripheries over time, the fringe areas not only inevitably lose their newness, but also tend to lose their distinction from the city; hence, they may stop being regarded as suburban, even though their residents may continue in a "suburban way of living" (Moos and Mendez 2015). In terms of these two basic characteristics of suburbia, socialist mass-housing definitely fits the bill. In the post-World War II decades, the mass-housing districts became the most distinctive new part of large Eastern-bloc cities, which enveloped the urban historic centres on all sides (Hirt and Stanilov 2009). In Russian cities, where socialism had arrived a quarter of a century earlier, *mikrorayoni* – albeit smaller than became the post-war norm – were first proposed in the 1920s. New Soviet cities for the new "Socialist Man" such as Magnitogorsk were often championed by Western, especially German, architects such as Ernst May, Mart Stam, Margarete Schütte-Lihotzky, Hannes Meyer, and Bruno Taut (the so-called May Brigade and the Bauhaus Brigade) prior to the time these architects had to leave the USSR in the 1930s. The famous General Plan of Moscow from 1935 displayed a series of large, new construction sites (*kvartalyi*) for 15 million m^2 of new housing that would have more than doubled the size of the old city by absorbing some 300 km^2 of agricultural land. World War II prevented the implementation of this plan, but all of its successors in the war's aftermath carried on the vision of massive peripheral expansion. The novelty of the socialist-era housing districts went well beyond the temporal. Their physical form – modern buildings arranged in neatly geometric super-blocks – represented an obvious rupture with the existing messy urban fabric, which was composed of small, irregularly shaped city blocks framed by narrow streets and eclectic, tightly lined-up, low-rise buildings (and, in the case of Russia, much of the existing, pre-revolutionary urban housing stock comprised shabby, single-storey wooden structures).

Several authors have noted that in its general layout, the Soviet mass-housing district, the *mikrorayon*, shared some essential physical features with the Anglo-American concepts of the "garden city" and the "neighborhood unit," both of which heavily influenced suburban design in Western countries throughout the twentieth century (e.g., Scott 2009; Dremaite 2010). Far from being a coincidence, the similarities resulted from the intense international exchange of planning models that did not dissipate with the formation of the Soviet Union and continued through the Cold War of the 1950s, 1960s, and 1970s (e.g., Dremaite 2010). Early Soviet proposals for decentralized urban growth, such as those by architect Nikolay Milyutin, were conceptually affiliated with the "garden city" movement and owed much to Arturo Soria's earlier ideas for a "linear city" (Kopp 1970). Like the Western concepts, the designs of Milyutin as well as those of other Soviet "disurbanists" from that period sought to abandon large cities. In the Soviet case not only were these cities detested for their crowding and pollution, much as they were in Western settings, but they carried the additional heavily ideological stigma of being products of the bourgeois order and serving as potential nodes of foreign-inspired counter-revolutionary plots. Much like their Western cousins, the Soviet disurbanists sought to decentre future urbanization patterns, provide new residential areas with generous common spaces and greenery, and treat them as separate but cohesively organized neighbourhood entities. Eventually, the radical disurbanist proposals were rejected by the Soviet top brass for being too expensive and impractical. Conveniently, about a decade after the formation of the USSR, all Soviet cities were declared already "socialist," thus removing their earlier stigma of being bourgeois remnants (Kaganovich 1931). But while the revolutionary disurbanist ideas faded, what eventually established itself as the dominant state-sponsored approach of urban restructuring – mass-housing districts around old urban centres – kept some of the disurbanist design philosophy alive, much as it kept the idea of strictly and comprehensively planned neighbourhood entities.

Certainly, there were clear contrasts between the Soviet and the Western neighbourhood design approaches. First, the socialist *mikrorayon* took on an increasingly monastic high-rise look, which was not as popular in North American Anglo-Saxon suburban settings, particularly in the United States (it was more popular in Canada). As industrialized construction methods became more advanced, buildings in the *mikrorayon* grew progressively larger over time, from an average of five storeys during the Khrushchev era to 10 to 15 storeys tall in the Brezhnev era. As a result, in sharp contrast to those of US metropolises, Russian urban densities increased sharply from city centre towards periphery. In Moscow, for example, they increased all the way to twenty miles away from the city's historic centre (Graybill and Mitchneck 2008). Eventually, the mass-housing districts achieved very high concentrations of population in all large Russian cities and, to a smaller extent, in large cities across the Eastern bloc. In this, they clearly failed to meet the third essential feature of suburban form proposed by Harris (2010): a sprawling, low-density pattern. Another clear difference between Soviet and Western mid-century peri-urban forms stems from the fact that, while Western planners working in the neighbourhood-unit tradition tended to focus on the design of the unit itself, their Soviet colleagues treated their version of the same concept as the building block of a larger unit, the *rayon* (region), which in turn was the building block of the urban whole – the comprehensively planned, modern city (Maxim 2009). Hence, unlike "typical" Western, especially US, suburbs, which were purposefully planned in isolation from the urban core, Soviet-style residential districts were supposed to be well integrated into the urban and metropolitan whole. Their street networks linked them to other nodes of urbanization, and they were widely served by public transit. In addition, they had a better planned functional balance. Shops, schools, hospitals, and other vital services were invariably shown in district plans in pedestrian or mass-transit proximity to the residential buildings. However, the benefits of connectivity, accessibility, and comprehensive service provision were often more a matter of socialist planning theory than of practice. Providing housing was a more vital and urgent need, since the right to housing was guaranteed in the Soviet and the other Eastern-bloc constitutions. But service provision lagged. According to some authors, in areas developed in the 1950s and 1960s the proportions of unbuilt services in urban mass-housing districts in Soviet cities reached 70–80 per cent for schools and day nurseries, 50–60 per cent for food shops, and 40–50 per cent for polyclinics (Krasheninnokov 2003). This scarcity forced the average Soviet citizen to cover long distances to reach what she or he needed, albeit by foot or mass transit rather than by car. The situation was even worse in small towns around large cities, which were also often dominated by standardized housing but were outside the official urban boundaries. These towns lacked not only services, but also jobs. Some million and a half residents of "dormitory" rural settlements and satellite towns around Moscow had to commute by train to the capital city in the 1970s (Rudolph and Brade 2005; Mason and Nigmatullina 2011). The inconveniences resulting from the poor integration of housing with jobs and services were a popular source of grievance across the Eastern bloc; in some countries people referred bitterly to their home districts as the "un-complex complexes" (Hirt 2012). The other widely shared citizen complaint, known to anyone with lay knowledge of Eastern-bloc cities, was the overwhelming dullness of the housing districts – an inherent feature of modernist design that in the socialist case was aggravated by the fact that architectural creativity was almost fully subdued to top-down political imperatives for the cheap production of housing (Lizon 1996; see Figures 2.1, 2.2, and 2.3). The dullness was perhaps best captured in the Soviet romantic comedy *The Irony of Fate*. This popular show from the Brezhnev years included scenes showing architects driven to desperation by know-all bureaucrats. The main hero in the show finds himself by chance in a city not his own. Yet the residential buildings look so similar to those in his home town that he does not recognize his mistake, which leads him to various personal complications. It is not too much of a stretch to see the mind-boggling repetitiveness of the Soviet housing towers made of perfectly identical units staggered over each other as the vertical, larger (and more depressing) version of Levittown.

Processes

Under the political economy of state socialism the process of conversion of rural to urban uses that enables the spread of (sub)urbanization was state-led. Similarly, housing was produced and distributed through a process wholly different from that used in free-market societies (e.g., see Andrusz 1984; Enyedi 1996; Szelenyi 1996). The difference was especially stark with reference to the United States, where, in contrast to some West European mid-twentieth-century welfare states, direct government provision of housing traditionally had been directed to only a very narrow, low-income segment of the population. As is widely known, the overwhelming majority of US dwellings have always been produced by the private sector on private land, converted from rural to urban land by private developers, and the government's role in housing has comprised mostly regulation, incentives, and infrastructure investment, notably highways.

Obviously, the Soviets took the opposite approach. In 1918 a Soviet decree abolished the private ownership of land. In the East European

Figure 2.1 Moscow, Russia. Photo: Alexander Slaev

Figure 2.2 Riga, Latvia. Photo: Danil Nagy

Figure 2.3 Belgrade, Serbia. Photo: Ute Lehrer

satellites, similar laws were passed in the late 1940s. Only in the more liberal socialist states, such as Poland and Yugoslavia, did a substantial percentage of land, mostly in agricultural areas, remain in private hands. In addition, while socialist laws permitted the so-called personal possession of goods (e.g., dwellings, cars, and all other household and consumer items), they banned the private ownership of large real estate and means of production, which, according to communist philosophy, led to exploitation of workers. Hence, all major industrial and agricultural enterprises, as well as the more sizable homes and residential estates, were nationalized a few years after the ascendancy of communist governments. In most countries, a large majority of the housing stock was taken over by the public sector. In Russia just a quarter of the housing stock remained privately owned by the close of the Soviet period, most of it in the form of single-storey homes located in small towns and rural areas. There were some exceptions. The relatively liberal regimes did not strictly follow this policy, and in some of the countries led by orthodox communist regimes, such as Bulgaria, the share of private ownership was high (in Bulgaria, over 80 per cent; Hirt 2012). But even in these cases the production, distribution, and maintenance of housing was a public responsibility, while the land under the residential buildings as well as the shared spaces within them were in public ownership.

The Soviet authorities consistently built new collective apartment buildings in and around cities throughout the socialist period, excluding the times of war, as millions of people moved from the countryside to fast-industrializing urban areas. The percentage of urban population out of the total population of Russia rose dramatically from 17 per cent to 73 per cent between 1926 and 1989 (Becker, Mendelson, and Banederskaya 2012). Urban mass-housing rose to top political priority, especially under Khrushchev, and remained so until the economic stagnation of late socialism. Under the program envisioned by Khrushchev, some 60 million people were relocated to standardized units in brand-new housing districts around urban centres within twenty years. The development process involved the appropriation of green and agricultural land within city limits and the designation of territories for the new housing districts by state planning offices in accordance with national and republican objectives for industrial, population, and housing growth set in the Soviet five-year national economic plans. The districts were comprehensively designed by state architectural offices and constructed by gigantic public homebuilding and infrastructure corporations. Finally, the dwellings were distributed among the individual households (either via sale or for rent, depending on the country), which was typically organized through the household head's place of employment. The vast scale of development would have been impossible without modern construction methods. These methods, much like the very idea of the comprehensively planned neighbourhood, were the offspring of intense international exchanges. The Soviets valued technological efficiency and economies of scale perhaps more than anyone else in the world. After all, they aimed to house millions of people without letting them inhabit informal, slum-like dwellings that would have compromised the regime's ideological commitment to equality and decent housing for all. Hence, they were willing to work with whatever was the most efficient building technology, regardless of its place of origin. The idea that prefabricated construction methods should become state policy was first endorsed in 1954 by Khrushchev in a famous speech on architecture and was turned into a decree by the Council of Ministers a few months later. The decisions of the ruling political apparatus were accompanied by a sequence of trips by the top Soviet architecture and engineering experts to several Western countries, including Britain and West Germany. The specific production and assembly technologies that were eventually selected and put into practice were a mix of Western and Soviet innovations (Dremaite 2010).

As prefabricated dwellings became the standard solution to housing in cities, private construction of individual houses within urban limits was strongly discouraged. Between the mid-1960s and mid-1980s, in fact, it was prohibited in all Soviet cities exceeding 100,000 inhabitants (Mason and Nigmatullina 2011). Building and ownership of individual houses was possible on the outskirts of cities in the form of small recreational cottages with garden plots – the famous Russian *dachi*, which by the end of the Soviet era were available to nearly all households residing in large cities (Becker, Mendelhson, and Banederskaya 2012). As permanent homes, individual units could be constructed typically only in rural areas and small towns around large cities. Homes in such peripheral locations were often coveted by people who wished to work in cities but, under the infamous Soviet *prospiska* system, could not obtain formal permits to establish urban residence. However, these dwellings were very far from being suburban dream homes. They were much shabbier than the units in the modern mass-housing projects; the public infrastructure of piped water, paved streets, sewers, and electricity surrounding them was in poor shape. Dwellings in the old areas of cities, left over from the detested pre-socialist era, were also typically of lower quality than those in the new mass-housing districts. Unless they were architecturally significant landmarks, historic buildings were often left to quietly decay. They were rarely upgraded and typically lacked modern (by Soviet standards) kitchen and sanitation equipment. This neglect was another tool that the Soviets used to systematically push populations towards their cities' new peripheries. Thus, if there was such a thing as a Soviet Dream (not a term in official use), by the 1970s it surely looked like a modern apartment unit in a mass-housing district. The dream may have been a bit dreary, and the apartments quite small, but they comprised the best housing alternative – which by the years of "mature socialism" was made available to vast segments of Soviet urban society, including most of the Soviet middle classes, from skilled workers to intelligentsia. The socialist state had used its immense powers to ensure that this was the case.

Philosophies

That housing policy and housing form were taken as serious expressions of political philosophy by the mid-century leaders of the two primary Cold-War rivals was made obvious to millions of American and Soviet TV viewers in the famous "kitchen debate" between Richard Nixon and Nikita Khrushchev. This debate took place in a model suburban house displayed at the American National Exhibition at Moscow's Sokolniki Park in July 1959. The US president clearly thought that the suburban model home – especially its kitchen – gave him an excellent opening to argue for the superiority of capitalism. He snippily emphasized how modern the kitchen equipment was and how affordable the home was to the American masses. The Soviet chief shot back, stating that his countrymen, too, had good access to equally modern, if not even more modern housing units. Nixon pointed out that, unlike their Soviet counterparts, Americans could select their house: "We have 1,000 builders, building 1,000 different houses … We don't have one decision made at the top by one government official." Standing his ground, a sharp-tongued Khrushchev emphasized that Soviet citizens had a constitutional right to housing: "In Russia, all you have to do to get a house is to be born in the Soviet Union. You are entitled to housing ... In America, if you don't have a dollar you have a right to choose between sleeping in a house or on the pavement." The debaters even used the kitchen to score points on women's rights, as they understood them. Nixon praised the modern kitchen for making the life of the American housewife easy and happy; Khrushchev responded that the Soviets disagreed with the "capitalistic attitude toward women" (*The Kitchen Debate – Transcript* 1959).

Khrushchev's pride in Soviet housing was deeply embedded in a long-standing Soviet belief that housing form, along with all other forms of architecture and urbanism, was an indispensable tool for the transformation of individuals, society, and culture (Crowley and Reid 2002). Hence the very earnest early Soviet debates on the nature of urbanism and its potentially problematic relationship to the old bourgeois order. Indeed, as some doctrinaire communists argued in the 1920s, how can a wholly new, communist society be built starting with the old, capitalist cities? As mentioned earlier, this debate was eventually put to rest in the early 1930s by a simple declaration that all Soviet cities already were "socialist." This declaration first came in the famous speech, and later a book titled *The Socialist Reconstruction of Moscow and Other Cities in the USSR* by "Iron" Lazar Kaganovich (1931), one of the top Soviet leaders of the 1930s and 1940s, who had *almost* become Secretary General of the Communist party after Lenin's death and had played the lead role in Moscow's planning and building activities in the decades preceding World War II. In Kaganovich-era propaganda, creating a new architecture, radically remodelling cities, and radically remodelling individuals and society were inseparable tasks: all were part of the total "socialist reconstruction of life and habits" (1931, 82). In the words of other Soviet classicists from the same period, such as El Lissitzky – one of the few lucky avant-garde architects who was not harmed by Stalin's purges – architecture and urbanism were "the material synthesis of culture" (cited by Baladin 1968). They had a key role to play in the total reformation of society and the creation of a new kind of citizen, the "socialist man," who was supposed to possess a vigorous sense of responsibility for the common good. This "socialist man" would emerge in "socialist cities" because there he would inevitably learn the values of selflessness and civic engagement by sharing space with his comrades. He would reside in collective housing structures and happily socialize with others in generously equipped workers' clubs. In Lissitzky's view, the "structure of our culture must be based on collectivist ideals." Architecture and urbanism had to both reflect and shape these ideals (cited by Baladin 1968).

The most radical Soviet proposals for a new type of residential architecture and a new type of city advocated the nearly complete elimination of private life. As explained in Milyutin's *Sotsgorod* (*The Socialist City*): "the ever-increasing drive toward collectivization of life impels us to build houses in an entirely different way than they have been up to this time and as they are still constructed in capitalist countries, where the basic economic unit is the family, each with its individual economy" (1975 [1930], 48). Both neighbourhood and dwelling design were means towards this end. Private family space was to be minimized by making the individual housing unit very small. This effort was likely driven as much by economics as by ideology in a country that could never achieve Western material standards, even though Soviet leaders perpetually claimed that it did, or would do so, in the foreseeable future. In contrast, collective spaces, from workers' clubs to neighbourhood shops, restaurants, parks, and gardens, were designed to be strikingly spacious. The even more revolutionary idea was not just to keep the dwellings small, but to take out of them most of the spaces designated for traditional (i.e., ostensibly bourgeois) family activities, such as cooking, washing, child-rearing, and guest entertainment. A progressively increasing number of these activities would be conducted outside the private home. Thus, in some cases dwellings were designed without kitchens, dining rooms, and/or bathrooms and all of these "services" were made available only for shared use. The extreme version of this idea was implemented in the infamous Soviet *kommunalki*, literally collective dwellings, where each family received a single room and shared all other facilities with several other families. In part, the idea of taking traditional family activities out of the private dwelling was driven by the Soviet belief that marriage and the nuclear family were capitalist institutions that led to the social entrapment of women. Children, from a very young age, were to be raised communally; cooking and cleaning were to be conducted collectively. This philosophy was behind Khrushchev's sneer at America's "capitalistic attitude toward women," as he put it in the "kitchen debate." From its inception communist ideology had strongly endorsed women's freedom. In Lenin's words (1966 [1920], 104): "Women's incipient social life and activities must be promoted, so that they can outgrow the narrowness of their philistine, individualistic psychology centred on home and family." Regrettably, the Soviets never achieved women's liberation, regardless of whether housing and neighbourhood design could ever have been the right tools to achieve it. And as early as the 1930s, more pragmatic and less revolutionary views were taking root in the Soviet Union. Just as cities were dubbed already "socialist" to avoid further debates on the total transformation of settlement patterns, the rhetoric for a minimalist, kitchen-free private dwelling was significantly toned down: "Today, when we have not yet built restaurants and cafes that entirely satisfy the requirements of the worker, the working woman and the working class family, it is impossible to resort to the wholesale abolition of the individual kitchen" (Kaganovich 1931, 85). Still, throughout the Soviet period, the ideals of collectivized lifestyles and reformed families persisted, if in weaker form. Propaganda for the "ideal communist city" written in the decades of "mature socialism" (1960s–1970s) still touted collectivist aspirations (Baburov and Gutnov 1971; Scott 2009), and the Soviet *mikrorayon* was forever marked by its modest private dwellings and its lavish public spaces.

It is easy to contrast the values that underpinned the Soviet push towards meagre private space and generous public space with those prevailing among American architects and urbanists of the mid-1900s. Take, for example, Milyutin's and Lissitzky's contemporary, the American icon Frank Lloyd Wright, whose proposals for Broadacre City came closest to becoming the blueprint for the mass suburbanization of the United States that occurred in the post-World War II period (Fishman 1982). Wright strongly believed in de-urbanization, to the point that he thought all cities larger than a county seat should vanish (a position similar to Milyutin's), but he held polar opposite, vehemently anticollectivist views about the balance of public and private space and the relationship between family and society. In Wright's opinion (1932), collective space had to be minimized and each American family awarded at least an acre to own and enjoy. In 180-degree opposition to the Soviet collectivists, Wright believed that the individual family was the building block of the society of the future: "We come now, to the most important unit in the [Broadacre] city, really the centre and the only centralization allowable. The individual home" (1932, 80). Still, this vivid contrast aside, the Milyutins and the Wrights of the 1930s shared a strong disdain towards old cities as being outdated and decadent – a sentiment in line with the views of most members of the Modern Movement. Another common conviction was that a certain type of urban and residential space was capable of building a certain type of social order, whether collectivist or competitive. This absolutist belief in the power of physical space was the province not only of architects but of mainstream politicians as well. Much as Kaganovich believed that spaces shape a world view, so did, say, President Herbert Hoover. In Hoover's view, the individual home was the "physical expression of [American] individualism, enterprise, of independence, and of the freedom of spirit" (cited by Archer 2005, 295). It stood at the "heart of our national well-being," it made for a "better citizenship" and served as a guarantee for American-style individual freedom (Hoover 1931, ¶3). The view was implicitly endorsed by mid-century Soviet-type town planners, who also taught that space crucially affects behaviour and feared that the "private yard" invites individualistic behaviour; that it "*makes* the bourgeoisie" (Hirt 2007a). The master builder of Levittown also saw the relationship between space and society in similar terms. In William Levitt's words: "no man who owns his house and lot can be a Communist; he has too much to do" (cited by Kelly 1993, 164).

Conclusion

The intense competition between American laissez-faire capitalism and Soviet state socialism brought the world to the brink of global war. As this chapter has argued, the competition extended to the realm of space: housing, urbanism, and architecture. The iconic urban peripheral housing districts that the two Cold War rivals produced in the middle of the twentieth century – the low-density American suburb and the high-density Soviet *mikrorayon* – looked very different, perhaps as different as one would expect, given the dissimilar approaches to land development and housing production and distribution that the two regimes took and their vastly different views on the relationship between individual and family on one side and society on the other.

Yet, contrary to conventional wisdom, beneath the obvious contrasts and the warring rhetoric that the regimes employed lurked some societal challenges, goals, and assumptions that were *not* worlds apart. As a number of political scientists and philosophers have argued, both the mid-century Soviet state and its American archrival sought to transform themselves into and present themselves to the outside world as quintessentially modern and democratic (as they saw it) societies that were destined to rule their age. Both wished to provide decent housing for most people. Both pursued an essentially modernist vision of progress and development by relying on a secular, science-minded bureaucracy and by embracing industrialization and Fordist/Taylorist methods of mass production (Murray 1992; Inglehart 1997). Hence the conclusion reached by philosophers of the rank of Vaclav Havel (1992) that state socialism and free-market capitalism were the two main versions of Western modernity. Most communist ideologues, say, Lenin – an earnest admirer of Henry Ford, who thought that Fordist methods should be adopted not just in the management of factories but in the management of whole states – likely would have agreed with this claim, while pointing out that their side was simply modernity's better half.

The iconic housing districts of the mid-twentieth century built on the two sides of the north Atlantic shared as much. Both were constructed in response to the challenge of a mid-century demographic boom, shrinking agricultural land, and growing industrial cities at a time when society came closer to providing equal opportunities in housing and elsewhere than ever before (and more than today). The visionaries behind them shared an expansionist attitude towards land and nature, a taste for novelty and progress, and a disdain for tradition and old cities. Both took Fordism and the latest technology from the factory to the countryside surrounding cities to produce mass-housing landscapes of a new kind. As the "kitchen debate" illustrated, regimes in both societies believed that their superiority could be proven, at least in part, by showcasing their particular version of a "spatial contract": the American side claimed to have created the conditions under which decent housing units were available to vast swathes of the US population in new suburban areas; the Soviets thought they had done even better by guaranteeing the right to modern housing to all people in expanded cities. Physical determinism was another shared trait typical of both regimes and of the wider Modern movement. In both cases building modern housing landscapes at the edge of cities was underpinned by a strong belief that space shapes society. Moreover, while favouring different kinds of spaces, the Soviets and the Americans implicitly agreed on what kind of spaces would make what kind of people. American ideology of space in the suburban era clearly favoured the individual and the family, expecting them to buy into the ideals of American capitalism by making generous private space widely available (e.g., Conzen 1996). And since it's capitalism that the Soviets feared, their newly created spaces were meant to serve as compressors of the private realm and amplifiers of the collective. Yet in fearing private space and seeking to minimize it, the Soviets implicitly but strongly concurred with their archrivals on the role that private space played in society.

From this point of view, then, mid-century Soviet and American mass housing reflected some common assumptions. It pursued some shared goals (reject old-style urbanism, expand, build new) and used some shared means (mass production, modern construction methods). Thus, if we follow Zygmunt Bauman's proposal that twentieth-century communism and capitalism were the "two legs on which modernity stood" (1995, 148), perhaps we could say that expansionist Soviet mid-twentieth-century mass housing and its sprawling American counterpart were the two legs on which suburbia stood.

ACKNOWLEDGMENT

This research has been partially supported by a grant issued by the Center of Excellence at the University of Sofia St Liment Ohridski. The opinions are those of the author and not of the Center of Excellence. Special thanks to Professor Konstantin Grozey and Ms Kristina Ferandiova.

REFERENCES

Andrusz, G. 1984. *Housing and Urban Development in the USSR*. Albany: SUNY Press. https://doi.org/10.1007/978-1-349-05218-9.

Archer, J. 2005. *Architecture and Suburbia: From English Villa to American Dream House, 1690–2000*. Minneapolis: University of Minnesota Press.

Baburov, A., and A. Gutnov. 1971. *The Ideal Communist City*. New York: Braziller.

Baladin, S. 1968. *Arhitekturnaia teoria El Lisickogo [The Architectural Theory of El Lissitzky]*. Accessed 8 July 2013. http://web.archive.org/web/20080510054907/http://novosibdom.ru/content/view/607/32/ [in Russian].

Bauman, Z. 1995. "Searching for a Center That Holds." In *Global Modernities*, ed. M. Featherstone, S. Lash, and R. Roberson, 140–54. Thousand Oaks, CA: Sage. https://doi.org/10.4135/9781446250563.n8.

Becker, C., S. Mendelson, and K. Banederskaya. 2012. *Russian Urbanization in the Soviet and Post-Soviet Eras*. New York: United Nations Population Fund.

Conzen, M. 1996. "The Moral Tenets of American Urban Form." In *Human Geography in North America: New Perspectives and Trends in Research*, ed. K. Frantz, 275–87. Innsbruck, Austria: Sebstverlag des Insituts für Geographie der Universität Innsbruck.

Crowley, D., and S. Reid, eds. 2002. *Socialist Spaces: Sites of Everyday Life in the Eastern Bloc*. Oxford: Berg.

Dremaite, M. 2010. "The (Post) Soviet Built Environment: Soviet-Western Relations in the Industrialized Mass Housing and Its Reflections in Soviet Lithuania." *Lithuanian Historical Studies* 15: 11–26.

Enyedi, G. 1996. "Urbanization under Socialism." In *Cities after Socialism: Urban and Regional Change and Conflict in Post-Socialist Societies*, ed. G. Andrusz, 100–18. Malden, MA: Blackwell. https://doi.org/10.1002/9780470712733.ch4.

Fishman, R. 1982. *Urban Utopias in the Twentieth Century: Ebenezer Howard, Frank Lloyd Wright and Le Corbusier*. Cambridge, MA: MIT Press.

Gornostayeva, G. 1991. "Suburbanization Problems in the USSR: The Case of Moscow." *Espace, Populations, Sociétés* 9 (2): 349–57. https://doi.org/10.3406/espos.1991.1474.

Graybill, J., and B. Mitchneck. 2008. "Cities of Russia." In *Cities of the World: World Regional Urban Development*, ed. S. Brunn, M. Hays-Mitchell, and D. Ziegler, 254–95. 4th ed. Lanham, MD: Rowman & Littlefield.

Harris, R. 2010. "Meaningful Types in a World of Suburbs." *Research in Urban Sociology* 10:15–47. https://doi.org/10.1108/S1047-0042(2010)0000010004.

Havel, V. 1992, "The End of the Modern Era: Excerpts from a Speech Delivered at the World Economic Forum in Davos, Switzerland on February 4, 2004." *New York Times*, 1 March, sec. 4, 15.

Hirt, S. 2007a. "The Compact versus the Dispersed City: History of Planning Ideas on Sofia's Urban Form." *Journal of Planning History* 6 (2): 138–65. https://doi.org/10.1177/1538513206301327.

Hirt, S. 2007b. "Suburbanizing Sofia: Characteristics of Post-Socialist Peri-urban Change." *Urban Geography* 28 (8): 755–80. https://doi.org/10.2747/0272-3638.28.8.755.

Hirt, S. 2012. *Iron Curtains: Gates, Suburbs and Privatization of Space in the Post-Socialist City*. Malden, MA: Wiley-Blackwell. https://doi.org/10.1002/9781118295922.

Hirt, S., and K. Stanilov. 2009. *Twenty Years of Transition: The Evolution of Urban Planning in Eastern Europe and the Former Soviet Union, 1989–2009*. Nairobi: UN-HABITAT.

Hoover, H. 1931. *Address to the White House Conference on Home Building and Home Ownership*. Accessed 23 November 2012. http://www.presidency.ucsb.edu/ws/index.php?pid=22927&st=home+ownership&st1.

Inglehart, R. 1997. *Modernization and Postmodernization: Cultural, Economic and Social Change in 43 Societies*. Princeton, NJ: Princeton University Press.

Kaganovich, L. 1931. *The Socialist Reconstruction of Moscow and Other Cities in the USSR*. Moscow: Cooperative Publishing Society of Foreign Workers in the USSR.

Kelly, B. 1993. *Expanding the American Dream: Building and Rebuilding Levittown*. Albany: State University of New York Press.

Kopp, A. 1970. *Town and Revolution: Soviet Architecture and City Planning, 1917–1935*. Trans. T. Burton. New York: Braziller.

Krasheninnokov, A. 2003. "The Case of Moscow, Russia," *UN Habitat and Development Planning Unit of the University College, Understanding Slums: Case Studies for the Global Report*.

Lenin, V. 1966 [1920]. *The Emancipation of Women: From the Writings of V.I. Lenin*. New York: International Publishers.

Lizon, P. 1996. "East Central Europe: The Unhappy Heritage of Communist Mass Housing." *Journal of Architectural Education* 50 (2): 104–14. https://doi.org/10.1080/10464883.1996.10734709.

Mason, R., and L. Nigmatullina. 2011. "Suburbanization and Sustainability in Metropolitan Moscow." *Geographical Review* 101 (3): 316–33. https://doi.org/10.1111/j.1931-0846.2011.00099.x.

Maxim, J. 2009. "Mass Housing and Collective Experience: On the Notion of Microraion in Romania in the 1950s and 1960s." *Journal of Architecture* 14 (1): 7–26. https://doi.org/10.1080/13602360802705155.

Milyutin, N. 1975 [1930]. *Nikolai Miliutin's Sotsgorod: The Problem of Building Socialist Cities*. Trans. A. Senkevich. Cambridge, MA: MIT Press.

Moos, M., and P. Mendez. 2015. "Suburban Ways of Living and the Geography of Income: How Homeownership, Single-Family Dwellings and Automobile Use Define the Metropolitan Social Space." *Urban Studies (Edinburgh)* 52 (10): 1864–82. https://doi.org/10.1177/0042098014538679.

Murray, R. 1992. "Fordism and Post-Fordism." In *A Post-Modern Reader*, ed. C. Jencks, 267–76. London: Academy Editions.

Rudolph, R., and I. Brade. 2005. "Moscow: Processes of Restructuring in the Post-Soviet Metropolitan Periphery." *Cities (London, England)* 22 (2): 135–50. https://doi.org/10.1016/j.cities.2005.01.005.

Scott, S. 2009. "The Ideal Soviet Suburb: Social Change through Urban Design," *Panorama*, 58–62.

Szelenyi, I. 1996. "Cities under Socialism – and after." In *Cities after Socialism: Urban and Regional Change and Conflict in Post-Socialist Societies*, ed. G. Andrusz, 286–317. Malden, MA: Blackwell. https://doi.org/10.1002/9780470712733.ch10.

Tammaru, T. 2001. "Suburban Growth and Suburbanization under Central Planning: The Case of Soviet Estonia." *Urban Studies (Edinburgh)* 38 (8): 1341–57. https://doi.org/10.1080/00420980120061061.

The Kitchen Debate – Transcript. 1959. Accessed 21 January 2014. https://www.cia.gov/library/docs/1959-07-24.pdf.

Wright, F.L. 1932. *The Disappearing City*. New York: Payson.

chapter three

Differentiated Landscapes of Suburban Property: Path Dependence and the Political Logics of Landownership Rules

ANDRÉ SORENSEN

We are living through a period of unprecedented expansion of urban population and an even greater expansion of urban area. During the next forty years the world's urban population is projected to almost double from 3.63 billion in 2011 to 6.25 billion by 2050 (UNDESA 2012, Table 3.1). Urban area has long grown even faster than population, and global urban area has been projected to double over the nineteen years to 2030 (Angel 2012, 169). This chapter focuses on the transformations of land and property produced by these processes of urbanization. By definition, all growth of urban area involves the conversion of land from lower-intensity rural or natural uses to higher-intensity urban uses. The starting point here is that during processes of urbanization, new property rights in land are created, in the sense that if a single 10-hectare field with one owner is subdivided into fifty new parcels, then for each parcel a new set of property rights in land is established, and those rights are different from the property rights formerly associated with the 10-hectare parcel. To sidestep for a moment the diversity and complexities of informal land development processes, the initial focus here is on the developed countries, where land development control, land-use planning, building codes, and land registry systems are in effect.

This chapter concentrates on the rules and institutional processes involved in the creation of such new sets of property rights in land. As the institutions that specify and define newly created property in particular jurisdictions evolve over time (Salet 2002; Buitelaar, Lagendijk, and Jacobs 2007), within particular jurisdictions there is commonly both a geographical differentiation of the property rights associated with particular parcels and a temporal differentiation of property produced at different times. I argue that in this regard the condominium is a valuable case, as it is a radically new form of property that has very different property rights from those associated with, for example, older forms of freehold ownership. The distinct logics of condominium property ownership are reviewed in part one. Here the point is that as only a few jurisdictions had created legal specifications for condominium property before the 1960s, and condominium is now legally defined in most countries, it provides a useful demonstration of the claim that the legal regimes that structure the creation of new property rights in land do evolve meaningfully over time. As the kinds of property that are created by processes of urbanization do change over time, and cities grow in physical area through processes of suburbanization, geographically and temporally differentiated landscapes of property will be produced in the suburbs. The suggestion here is that this understanding generates important insights for understanding suburbs, and it generates productive research questions.

I argue that the emergence of these new, increasingly differentiated landscapes of property has been one of the least visible aspects of suburban development over the last fifty years. And condominium is just the most obvious change. Whereas in the unplanned and largely unregulated cities and suburbs of a century ago most land had similar property rights and most land uses and building types could be located anywhere, over the twentieth century the spread of planning regulatory frameworks, including land-use regulations and zoning, building codes, and new development control systems, produced an ever greater differentiation of urban space and of the property rights of the parcels created. I suggest that this differentiation of property in growing cities has profound and long-lasting consequences for people's relationships to place, homeownership, urban infrastructure and finance, municipal governance, and local democracy, and that these new geographies of property are an increasingly fundamental aspect of suburbanization processes that should be better understood.

It is also clear that in various countries the legal regimes structuring the production of new property are quite different, and that differentiation of the legal specification of property among and within jurisdictions is therefore an important window for the comparative study of suburbanization. This chapter seeks to highlight the insights gained by looking at suburbanization from the perspective of its transformations of land and property rights, the condominium being a particularly revealing case of a larger phenomenon.

It is surprising that so little research on the suburbs focuses on suburbanization as a process of creating differentiated property rights in land. There is a large body of research on urban property, from work at the macro level on the role of capital investment in real property as a key circuit of capital (see Harvey 1982) and on the financialization of urban property markets (Weber 2010; Rolnik 2013). At the micro level research has focused on the roles of municipal actors, developers, and development industry in the production of property (Weiss 1987; Healey 1992a, 1992b; Adams 1994; Fainstein 1994). Land policy has long been seen as a way of influencing development patterns (Darin-Drabkin 1977; McAuslan 1987; DeGrove and Miness 1992), but, as has repeatedly been pointed out (Krueckeberg 1995; Jacobs and Paulsen 2009; Adams and Tiesdell 2010), surprisingly little attention has been paid to the ways in which evolving legal frameworks of property law shape the production of property in growing cities.

Blomley's work on law, legal geographies, and landscapes of property (1994, 1998) sets out some of the groundwork for such research, and subsequent work by legal geographers examines theoretical and conceptual approaches to the role of law in differentiating and giving meaning to urban space (Blomley, Delaney, and Ford 2001; Holder and Harrison 2003; Taylor 2006; Valverde 2012). Blomley, in particular, has opened a number of productive approaches to the study of the spatial impacts of legal regimes and interpretations of property, particularly in his work on spatializing the concept of property and on the contested meanings of ownership (Blomley 1998, 2004). Williamson's provocative question "Is sprawl conducive to democratic citizenship?" links urban form and the production of new urban space with political theory and questions of social justice (Williamson 2010, 10). But none of this work focuses on the political and social implications of the evolving legal institutions that structure the production of new property rights in land during processes of suburbanization, even though the conversion of non-urban land to urban property is one of the most fundamental elements of the urbanization process, during which new properties, property rights, and claims are created.

States play a central role in regulating the production of new urban property rights during processes of land development and in defining and protecting the property rights that are created in those processes. To a greater degree than in perhaps any other market, the state is fundamental and indispensable to the existence and continued value of urban property and to the working of property markets.

Property Law 101 textbooks define property as consisting of bundles of property rights where different sets of rights attach to different properties. The concept of differentiated bundles of property rights makes it clear that there are multiple modes of property ownership, and that different properties can have very different "bundles" of rights. These varied new "bundles" of property rights are structured by development regulations, permit processes, and the legal systems that define and protect ownership.

For example, a property zoned single-family detached residential in a zone of similar properties has a very different bundle of property rights than a brownfield site in a redevelopment area, or a unit in a high-rise condominium, or an inner-city townhouse next to a gas station. Urban property values and meanings are profoundly influenced by surrounding properties and public goods (Fischel 2001; Fennell 2009). Property rights are therefore shaped by larger patterns of neighbourhood design, land use, infrastructure, and the sets of restrictions on land-use change at the neighbourhood and city scale that provide the structure for and constraints on the bundles of rights of individual properties. For example, Nelson (2005) describes zoning as a "collective property right" that compensates for the removal of individual homeowners' rights to develop whatever they wish with collective rights to a stable neighbourhood environment. The legal structure and regulation of property markets vary greatly between jurisdictions, as do the systems designed to put conditions and requirements on the production of new property (Booth 1996; Needham 2006).

I suggest that the example of condominium is revealing of important political and social issues associated with the production of property in processes of suburbanization, and it helps to make sense of the three main points that form the core of this chapter.

Part one develops the argument that condominium is a fundamentally different form of property ownership from earlier models of freehold and leasehold, generates different political logics and incentives, and has profound implications for our understanding of suburbanization. Recent literature on condominium suggests that far from being relatively homogeneous, property rights and configurations are increasingly differentiated in ways that have significant impacts on people's relationships to place, homeownership, and local governance processes. Rules regulating land development, associated infrastructure systems, public spaces, and the institutional structures designed to maintain newly created urban spaces are generative of and structure relationships between property owners and the local state and with their neighbours. Condominium property illustrates this proposition clearly.

The distinctive characteristic of condominium property is that ownership consists of two parts. Condominium means individual ownership of particular units, shared ownership of common spaces and land, and shared responsibility to manage and pay for maintenance of the shared spaces. In different countries and jurisdictions this form of property is also referred to as strata property, common-interest developments, gated communities, and private communities. Here I refer to all of them with the generic term "condominium property." This new form of property has characteristics quite different from traditional freehold forms of residential property – whether single-family detached, semi-detached, or townhouse – in which homeowners own both the dwelling and the land it occupies as a single unit of property. Condominium forms of ownership where individuals own their dwelling space and share ownership of land and common facilities are increasingly common worldwide, and the literature on condominium suggests that these new forms of ownership and property rights have significant political and social impacts, which are reviewed in part one.

Part two suggests that particular sets of property rights are relatively enduring, especially in the case of residential property where there are strong incentives to ensure continuity and prevent change, in order to protect asset values. I argue that the new urban property – whether on the urban fringe or as redevelopment – that is created during any particular period tends to bear a lasting imprint of the regulatory system that existed at the moment of its creation. Time- and place-specific landscapes of property, infrastructure, and regulation are produced at different times and in different places. This imprint is registered not only in the pattern of property divisions and of public and private land parcels, which are often enduring, but also in the title deeds and the specific bundles of property rights associated with the parcels created. Condominium is a particularly clear example of an institutional design that reinforces continuity over time, creating multiple barriers to incremental change.

The significant point here is that if we accept the argument in part one that over time the specification of property has changed in meaningful ways, and if we also agree that, once established, individual properties and sets of property tend to demonstrate significant continuity over time, then we generate a picture of unfolding landscapes of suburban property that are differentiated both spatially and temporally, depending on when particular properties and sets of property were created. The suggestion is that when new suburban land is developed, new units of property are created according to a menu that allows a range of different options, but that menu changes over time and is differently structured in particular cities and countries. The particular development control and property rights regime that exists at the moment of land development, as well as the creation of new property, leaves lasting imprints on landscapes of urban property in the form of specific sets of property that are specified in particular ways. Although cities and urban property landscapes are constantly changing through either redevelopment or changes to the rules, the property specifications created during subdivision tend to have long-lasting impacts, and condominium is particularly resistant to change because of institutional designs that favour continuity. These include supermajority voting rules to be able to change bylaws and management rules, behaviour rules, and strict rules against incremental renovation and change to units. In contrast, the owners of non-condominium freehold property can engage in a wide range of maintenance or renovation or even complete rebuilding with few restrictions.

In part three I suggest that if different forms and configurations of property have significant impacts on the incentives and interests of property owners, as suggested in part one, and some forms of property are particularly resistant to future incremental change by redevelopment and resale, as suggested in part two, then the evolving regulatory regimes that specify land development requirements and property products and the particular landscapes of property created must be central elements in the study of suburbanization processes. Understanding these new property landscapes is particularly important if we wish to compare suburbanization in different cities and countries, as contingent moments of legislative and political development create varied legal regimes of property in different countries, which themselves are likely to be highly path dependent.

This approach provides one answer to recent calls for new cross-national comparative urban research methods and theory that can move beyond the comparison of cities of similar size, wealth, and governance structure, to include the huge diversity of cities in developing countries, beyond the usual suspects examined by global cities research (Robinson 2006, 2011; McFarlane 2010). If we compare institutions of property, the property regime that produces them, and the landscapes of property that are produced, then we can investigate very different cities, since we are comparing not cities as a whole, but the institutional structures that shape them.

A brief concluding section then brings together the main implications of this set of arguments.

Differentiated Property Ownership Is Transformative of Suburban Space

The suggestion here is that suburban space is increasingly differentiated, not just by the well-known and visible differences of land use, socio-economic status, housing type, and infrastructure quality, but also by increasingly differentiated geographies of property. Such geographies are largely invisible, as the built forms of different sorts of suburban residential property can be quite similar. Yet the meanings of various property types, their production, regulation, and the behaviours and attitudes they give rise to can be profoundly different. Condominium is useful as a particularly clear example of the increasing differentiation of property rights in suburban areas and of the contingent and variegated roles of states in regulating urban property markets. But the suggestion here is that condominium is merely an extreme case of a much more pervasive phenomenon.

The leader in condominium development is the United States, where it has captured a large share of the market, producing over a third of all housing units since 1970 and by 2002 over half of all new units (McKenzie 2003; Nelson 2005). As of 2011 there were 314,200 condominium associations, with 25.1 million housing units, and 62.3 million residents in the United States (Community Associations Institute 2012). In the fast-growing sunbelt states of Florida, Arizona, Nevada, and California, in particular, a significant majority of new suburban development is condominium. Partly as a result, much of the existing literature focuses on the United States. However, similar trends are occurring in several other countries, especially in China, where a great quantity of new development is supplied as "residential clubs" with characteristics similar to those of condominium: private housing units and jointly owned and maintained neighbourhood amenities (Webster 2002; Lee and Webster 2006).

Webster, Glasze, and Frantz (2002) suggest that quite varied incentives exist for the emergence of these forms of property in different places. In post-apartheid South Africa the issue is security from crime (Jürgens and Gnad 2002), as it is in Latin America (Coy and Pöhler 2002; Coy 2006). In Britain the issue is rehabilitation of distressed social housing estates (Webster 2001; Blandy 2006). In Jakarta the motivation is insulation of the affluent from the poor (Silver 2008). Webster, Glasze, and Frantz (2002) are also careful to point out that although clearly influenced by US practice, different legal institutions and management rules have developed in each country, so the design, property specifications, and meanings of condominium property is varied.

The intent here is not to make normative claims either for or against these new forms of property, but instead to review recent debates about condominium to gain insight into distinctive features of these new forms of property. The focus is on those characteristics that serve to differentiate suburban spaces, generating and perpetuating different logics of their creation and use, and the reproduction of alternative property logics. Major issues that have been identified concern the institutional logics of the privatization of public space and associated loss of democratic space and civic engagement (Davis 1990; Low 2003; Mitchell 2003; Kohn 2004), logics of governance fragmentation and segregation (McKenzie 1994; Low and Smith 2006), logics of infrastructure provision, public goods, and municipal governance (Fischel 2001; Graham and Marvin 2001; Webster 2002; Nelson 2005), and the logics of surveillance and securitization (Davis 1990; Blakely and Snyder 1997; Sorkin 2008). Each is reviewed briefly.

One of the most prominent concerns addressed in the existing literature is the loss of democratic spaces and the "public realm" and associated constraints on civic engagement in suburban environments generally, and private communities particularly. Many have seen the emergence of gated communities and condominium as a clear expression of neoliberalization, linking privatization, the decline of public services, and the hollowing out of the local state (Low 2003; Mitchell 2003). Davis (1990) examined the links between privatization of public space, exclusion of unwanted populations, and increased security and surveillance in Los Angeles. Kohn (2004) takes this argument further, showing that a major consequence of the privatization of public space in US suburbs has been a loss of democratic space. Kohn argues that, when private spaces replace public gathering places, the opportunities for political conversation are diminished. Such political conversation includes "street speaking, demonstrations, picketing, leafletting, and petitioning. The face-to-face politics that takes place in public places requires no resources except perseverance and energy" (Kohn 2004, 6). Kohn shows that, even more than the enclosure of space documented by Davis (1990), the routine failure to create new public space as cities grow is a major constraint on democratic conversation. Neither Davis nor Kohn focuses primarily on condominium property, but the private ownership of shared condominium spaces, both inside as corridors and common rooms and outside as shared recreational facilities, green space, and road systems is probably the most important example of the growth of private space at the expense of public space in contemporary processes of suburban development.

Drawing on Sennett (1977) and others, Kohn also argues that it is essential for citizens of a democratic society to encounter in public space the full range of members of society, and that self-segregation into narrow subgroups inhibits social cohesion. A prominent claim is thus that private communities generate powerful logics of fragmentation and segregation that serve to increase hostility and suspicion of others (McKenzie 1994; Low and Smith 2006; Dredge and Coiacetto 2011). Kirby questions this distinction between public and private as overly simplistic, suggesting that many of the neighbouring behaviours and uses of privately owned public spaces in condominiums are similar to those in traditional neighbourhoods (Kirby 2008). He suggests that dire predictions that the privatization of public space will lead to a collapse of democracy and the diminishment of civic behaviour may be overstated. But even if neighbouring behaviours are highly prevalent within condominium developments, they are by definition separate from broader public engagements, as only condominium owners are a part of condominium neighbourhoods. Condominium and gated communities unavoidably divide and fragment political spaces of suburbs. And it is clear that voluntary activity on condominium boards and committees has very different motivations and structures than similar activities on school boards, community groups, or local environmental groups, as condo members as co-owners benefit directly from any improvements to or protection of their assets, whereas for regular community members such benefits are conceivable, but are likely to be much more diffuse.

Perhaps most troubling are the logics of surveillance and security (Davis 1990; Blakely and Snyder 1997; Sorkin 2008). Here the argument is that as people increasingly withdraw into private realms, the sense of risk associated with public space also grows. Increased security can induce heightened paranoia rather than a feeling of safety, and it influences behaviour (Caldeira 1996). Public streets are abandoned by those with resources and are left to the poor and the homeless, who are excluded from condominiums and gated communities. As Caldeira puts it: "In cities fragmented by fortified enclaves, it is difficult to maintain the principles of openness and free circulation which have been among the most significant organizing values of modern cities. As a consequence, the character of public space and of citizens' participation in public life changes" (1996, 303).

The argument is that this sense of separation from a broader public sphere is corrosive of a shared sense of public responsibility. While perhaps most extreme in gated communities, it is also likely in condominiums more generally, as their "public spaces" are actually shared privately owned spaces, not true public spaces. Condominium owners have diminished incentives to try to strengthen the capacity of public bodies to maintain the security and accessibility of public spaces and facilities for all.

Another decisive characteristic of condominium is that it provides an alternative mechanism for the finance, delivery, maintenance, and policy-making about community infrastructure. Condominium therefore generates different logics of infrastructure provision, public goods, and municipal governance than freehold property. Particularly in the case of suburban condominiums and private communities, which routinely provide amenities such as swimming pools, water supply and sewers, community meeting spaces, and security, co-owners have incentives to object to paying municipal taxes for similar services that are provided publicly by the local municipality (Fischel 2001; Graham and Marvin, 2001; Webster 2002; Nelson 2005). Unsurprisingly, radically different interpretations of this issue exist. Much of the literature describes condominiums as clubs that deliver shared amenities to like-minded residents, and is extremely positive about their benefits in providing specific sets of amenities to residents. The suggestion is that condominium can be much more efficient and responsive to individual preferences than can municipal governments, which tend towards a lowest-common-denominator approach to service provision (Fischel 2001; Webster and Lai 2003; Nelson 2005; Lee and Webster 2006).

Municipal governments often welcome condominiums, as they supply most of their own infrastructure, thus lessening the financial burdens of suburban growth. But although this aspect reduces some of the upfront costs of growth for municipalities, condominiums usually continue to depend on local governments for some key infrastructure such as schools, roads, and ambulance services, and they shed externalities onto local governments, so the actual balance of costs and benefits is less clear (Warner 2011). Others see the privatization of infrastructure provision as generating serious social equity problems, and a "splintering city" (Graham and Marvin 2001) in which those who cannot buy in to the new model increasingly lose out. Condominium developments clearly generate very different logics of public and private space, of civic engagement and democratic governance than earlier forms of freehold property did. They also create different logics and mechanisms for the finance, delivery, maintenance, and policy-making about community infrastructure and are likely to support different attitudes and interpretations of municipal public service provision and the value of contributions to local taxes.

The logic and legal basis of local governance is also increasingly differentiated between new private communities and traditional city and suburban municipal governments. Whereas the latter are governed by ideals of one-person/one-vote and the principle of equal access to services and are usually prohibited from discriminating among residents, private communities follow primarily market logics, are registered as business corporations, membership is compulsory with ownership, and opportunities to influence policy are limited. The main option for those who are dissatisfied is "exit" or moving out of the development, rather than "voice" or attempts to change policies. Legal constraints on behaviour are also quite different for businesses than for local government. For example, condominiums can legally discriminate based on age, marital status, and wealth, whereas municipalities cannot.

Finally, the logics of homeownership and community citizenship may be quite different in private communities from those in freehold ownership in traditional municipalities. While both are likely to share a tendency to not-in-my-backyard (NIMBY)-type opposition to locally unwanted land uses (LULUs), other aspects of homeownership are likely to be quite different. For example, the mode of defining community insiders and outsiders is by shared ownership of community assets in condominiums, whereas in traditional communities citizenship is shared by anyone who lives in, works in, or owns assets in a place. Ownership, rather than residence, becomes central to defining belonging in condominiums. And privatization and fragmentation of urban space can create significant management and governance problems. For example, McKenzie (1994, 2003) examines shortcomings of condominium regulatory frameworks that have resulted in high levels of litigation, weak and/or corrupt management, and failures to budget for major maintenance. He blames the fact that in many condominiums untrained volunteers are carrying out functions that used to be the responsibility of municipal governments. While weak and/or corrupt management also has been a problem for municipal governments, the approaches to solving such problems are quite different in condos and municipalities, thus supporting the claim of different political logics generated by this property model.

The impacts of these divergences in forms of ownership may be quite different when condominium is the dominant form of development compared with when it consists merely of scattered enclaves. It seems likely that there will be significant impacts on the functioning of the local state, public facilities, taxes, and services, if more and more of the public realm is actually private, and public resources consequently decline. This remains a little researched issue (Dredge and Coiacetto 2011).

If we understand the process of suburban development as one of creating new property rights, new patterns of public and private space, new patterns of infrastructure, and new revenue streams to support them, then the emergence of condominium as a new form of property that generates different kinds of ownership and very different ownership patterns, behaviours, and governance logics is clearly a significant change that should be better understood. And the larger probability that different configurations, designs, densities, and landscapes of property will lead to varied political logics of place must be considered, even though condominium may be a particularly dramatic case.

Path Dependence and the Institutionalization of Property Rights

A key argument advanced in this paper is that condominium ownership structures create new and particularly path-dependent institutions and spatial patterns, and that such path dependence reinforces the long-term impacts of the moments of property creation (see Sorensen 2015). There are two main reasons why condominium is a highly path-dependent ownership form. The first, and most obvious, is that land development and subdivision are always inherently costly to reverse. The second is that, in order to protect the long-term value of the property, condominiums create complex ownership structures that are intentionally designed to be hard to alter. This does not mean that no change will take place, but it does suggest that some types of change will be much more likely than others.

The first point is that land subdivision and development in itself is a highly path-dependent process. Land subdivision is is characterized by the transformation from less to more fragmented property ownership; from less to denser, more complex patterns of institutionalization and legal protections; and from lower to higher property values. The urban process with the greatest degree of path dependence may be the subdivision of land and the creation of new urban parcels, housing units, streets, public infrastructure, parks, and other urban spaces. The grids laid out by the Romans all over Europe 2,000 years ago are still imprinted on European cityscapes, as are the footprints of their fora, stadia, and baths, even where the actual buildings have long since disappeared (Kostof 1991). Property divisions remain even when bombing, fires, or earthquakes raze all the buildings in urban areas. And while over time one might see continuity of a particular building or aqueduct as resulting from factors more literally concrete and less political than, say, a specific public policy, in fact property divisions and the protections offered them by states are unquestionably political and institutional in nature.

That is, when a developer subdivides land, builds houses, and sells them, what is being sold is not merely a piece of land and a building. It is actually a much more complex product, which includes the promises implicit in the residential zoning that an auto wrecker or nightclub will not be located next door. It also includes a promise that the street in front will be maintained; that water supply and sewers will work; and that local schools will be provided, crime prosecuted, and property rights protected.

Seen this way, it becomes clear that urban property is embedded in an extraordinarily complex mesh of interlinked rights, obligations, guarantees, and risks. Some are formally recognized in title deeds, some are contracts between property owners and municipal governments, some are unwritten and customary, while others are generated merely by proximity (see Fischel 2001; Fennell 2009). All can have profound impacts on property resale values, and many are highly path dependent because individuals have powerful incentives to fight against changes that they think will reduce property values. Such changes are therefore often contested by NIMBY movements. Such "defensive localism" is a fundamental driver of urban politics (Dear 1992; Walsh, Warland, and Smith 1993; Cox 2002). However, the point is that in this case path dependence and the positive feedback effects ensuring continuity derive not from the sunk investment in and enduringness of buildings, but from the motivations and values of their owners and users. We should therefore see the development of new suburban areas not merely as a construction process, but more as a highly complex process of urban institutionalization in which dense new networks of rights, obligations, relationships, and expectations are established.

In fundamental ways, therefore, the process of urbanization is a path-dependent process. When farm fields are turned into subdivisions, it is virtually impossible to turn them back into farm fields. This is not primarily because the topsoil has been stripped, trees cleared, contours levelled, roads paved, and pipes installed. Such changes conceivably could be reversed. Much harder to undo is the institutional mesh of property rights, expectations, and obligations, as well as the vastly higher property values post-development. So, while the enduring image of suburban subdivision development is of bulldozers clearing the topsoil and the construction of roads and houses, they are not really the decisive acts that create a new suburban reality and forever displace the prior rural setting. Rather, it is the private and public act of creating and guaranteeing new sets of property ownership institutions and relationships and the higher property values they protect that is a generator of path dependent institutions.

The institutional configurations created in processes of urbanization have particularly high levels of path dependence in North America because over the twentieth century the practice of residential planning embraced the imperative of creating physical and institutional designs that were explicitly designed to resist incremental change (see Figure 3.1). This was both a response to the emotions driving defensive localism and a reinforcement and legitimation of them. The fear of blight, slums, mixed uses, visible minorities, and declining property values provided powerful incentives for single-use residential developments with highly restrictive zoning, covenants and deed restrictions, minimum lot sizes, winding and looping roads that block through traffic, green buffers insulating houses from different land uses, and other such strategies (Goldsmith 1995; Weir 1995; Sies 1997; Hayward 2009). As Lasner (2012) has shown, in the United States condominium ownership was in part motivated by the desire to maintain exclusion policies even after the US Supreme Court outlawed racially restrictive zoning and covenants.

Second, among urban property institutions perhaps the most path dependent is condominium. Condominium creates ownership deeds with shared ownership of common facilities, shared obligations to pay

Figure 3.1 Building boom in Toronto, Canada, suburbs. Building booms create extensive new sets of urban property that are enduringly imprinted with the characteristics and meanings of the institutions used to create them and the values of the actors driving these processes. Photo: André Sorensen 2002

ongoing maintenance and administrative fees, and private ownership of individual housing units. Shared ownership and management of condominium property is regulated by bylaws and "Covenants, Conditions and Restrictions" (CC&Rs), creating a kind of "neighbourhood constitution" (Nelson 2005) that can be changed only by supermajorities of owners. This format creates a highly path-dependent set of institutions that not only are explicitly designed to resist significant change, but also provide enforcement mechanisms to enable the surveillance and punishment of those attempting incremental changes to their own unit or behaving in a manner defined as deviant.

Such ownership forms can be particularly powerful because co-owners are educated to understand individual behaviours, idiosyncrasies, and maintenance approaches to be a threat to collective property values. Individual owners therefore have powerful incentives to see their self-interest as represented by continuity, and they will be inclined to resist change. Further, upon purchase, condominium owners contract with the Association to cede individual rights to alter their property except in specified ways, and implicitly and/or explicitly agree to police the behaviour of other owners. Restrictive CC&Rs permanently embedded in title deeds and ownership contracts regulate a wide variety of behaviour (Warner 2011, 159). Developers create most of these rules prior to property sales and usually maintain control over developments until all units are sold, and supermajority votes of co-owners are required in order to change most significant elements of the rules. This mechanism ensures that purchasers self-select to be those who are supportive of the status quo.

The gated community, condominium, or "common interest development" is thus one in which the coercive powers of the state are voluntarily subcontracted to a private body and in which defensive localism is enshrined in physical, institutional, and legal design (see Figure 3.2). This shift from public enforcement of property rights and obligations to private enforcement has the paradoxical side-effect of creating a much stronger bias towards continuity and stronger enforcement mechanisms. Such developments are likely to be particularly path dependent compared with traditional freehold and leasehold property forms, which are much more differentiated and more permissive of incremental change. For all these reasons the growing share of condominium property in suburban development can be expected to produce a significant and enduring impact on local political life, attitudes, and outcomes.

But again, it is important to consider the extent to which condominium is simply the clearest case of a much more pervasive phenomenon. The carefully planned affluent suburban neighbourhood, with minimum lot sizes, restrictive zoning bylaws that allow only detached single-family homes above a certain size, lawn maintenance bylaws, and mandatory insurance coverage produces some similar logics of ownership, behaviour, and continuity as those described for condominium. The processes of institutionalization of new urban areas, property rights, and public and private infrastructures in new suburbs do in many cases create path-dependent institutional configurations that are resistant to change, but are not unchanging. It is not necessary to claim that these configurations are good or bad to suggest that they are important.

Figure 3.2 Privately governed housing developments with shared ownership of common facilities such as these outside Las Vegas have become a major component of suburban development around the world and have profound implications for people's relationships to place, homeownership, community, municipal governance, and local democracy by creating new forms of suburban property ownership with different political, property, infrastructure, and social logics than traditional freehold and leasehold forms.
Photo: André Sorensen 2007

Systems Governing the Creation of Property Rights

If the forms and configurations of property influence political incentives and behaviours, impact patterns of neighbouring and social relations, and are likely to be enduring, then the regulatory and legal systems that structure the production of new property rights become central to the comparative study of processes of suburbanization, both within and between countries. Condominium provides a useful example, as it produces a radically new form of property, with significant political and social impacts. In fact, however, this fundamental point applies more broadly to all aspects of systems that produce new property rights and to all forms of urban property.

States create legal and planning systems to regulate the production of new urban property. These systems define the varied bundles of rights associated with individual properties, permit certain forms of land subdivision and property creation and prohibit others, and record ownership through land registries and cadastral systems. They also define minimum infrastructure necessary, and design, finance, and manage much of the basic infrastructure and services necessary for urban and suburban space to be viable. As suggested above, the property produced by these systems is usually enduring, producing very long-lasting facts on the ground in the form of new configurations of property, new public and private spaces, new bundles of property rights, and the debt obligations and revenue streams associated with a variety of infrastructures. Certainly, some of these facts change incrementally over time, but the imprint of the system-regulating processes of land division and property creation in suburbs, structured by its time- and place-specific menu of property options and specific restrictions and requirements that permit the creation and registration of new property, is all but indelible.

Evolving approaches to planning and the regulation of urban development over the course of the twentieth century meant that systems regulating the production of property and the geographies of property created at different times outside any given city have also changed over time, creating differentiated landscapes of property that are varied both across geographical space and by time of development. Path-dependent trajectories of institutional development in different countries mean that systems regulating the creation of new property vary greatly, and the forms and specifications of the property they produce are also extremely varied, even when the built form of buildings is superficially similar. For example, a condominium property built in Florida in 1970s has very different property rights from those of one built in 2010, and that property in turn has very different property rights from those of one built in the same year in Ontario or in Japan, because of differing legal systems and traditions.

Condominium law in Japan, Florida, and Ontario is instructive, as even though condominium laws in these three jurisdictions were enacted within a few years of each other – in 1962, 1963, and 1967, respectively – initially they had significant differences and subsequently have followed different evolutionary pathways. A distinctive feature of the Japanese law is that it followed the practice of the existing Civil

Code regulations for shared ownership and required agreement of 100 per cent of owners before changes could be made to bylaws or major renovations could be undertaken. This meant that repairs were often delayed and decision-making was difficult, which has greatly reduced the resale value of condominiums in Japan. A major goal of legal reform in 1983 was to revise the rule for changes to bylaws and major maintenance from 100 per cent to a supermajority of 75 per cent.

Also, Japanese property law recognizes ownership of buildings and the underlying land as separate issues; since this distinction was not adjusted for condominium, condo owners had three forms of ownership: their unit, a share of common space and facilities, and a share of the underlying land. In several cases the ownership of the underlying land became separated from the unit ownership, causing significant problems for decision-making and subsequent sales. Legal systems for property clearly have significant differences in different jurisdictions, which can have long-term impacts on property ownership, including of condominium.

Japanese condominium law is also a national law in a country with a civil code legal system, whereas both Florida and Ontario are subnational jurisdictions within federal states that created their own condominium laws and planning systems, and both share the English Common Law as the basis of their legal systems. In civil code legal systems court cases and precedents have little impact on legal interpretation compared with results in common law systems, and legal change comes about almost entirely through legislative revision. In both Florida and Ontario large numbers of court cases not only contributed directly to legal change, but also put pressure on legislatures to change the laws. In Japan going to court to solve disputes is rare, partly because it is so expensive (Upham 1987; West 2005). So, processes of legal evolution have been quite different in the three jurisdictions.

The Florida case points to other factors that increase the differentiation of legal property regimes. Enacted at the height of a real estate boom in a state where property developers were major political players, the law was relatively weak in protecting the interests of homebuyers and was enacted primarily to enable new forms and configurations of development (McCaughan 1964–5). The act failed to stipulate rules of management agreements or bylaws for the homeowner associations that manage condominiums. Transparency of contracts for maintenance and property management was very limited, and developers exploited these weaknesses by establishing long-term contracts with built-in escalator clauses and penalties for cancelling contracts. In some cases units were sold citing artificially low maintenance fees, which resulted in eventual budget crises and imposition of drastically higher fees. Such problems led to large numbers of court cases and eventually to calls for increased regulatory oversight and for revisions to the law (Poliakoff 1992, 474). One consequence is that legal revision in Florida has been frequent and sometimes chaotic. A major reform in the 1970s was necessary because the many minor revisions and court cases had introduced contradictions and inconsistencies into the legal regime. Condominium law was seldom the subject of national political attention in Japan and legal revision was infrequent, whereas it was often highly contentious in Florida, where disgruntled unit holders had much closer access to their legislators, with the result that legislative change to condo law happened almost every year from its initial passage up to the 1990s (ibid., 473).

Two further elements of differentiation worth mentioning are timing and governance system. Ontario's law was much less contentious than that of Florida, in part because it was enacted later and those drafting the Ontario law were aware of large numbers of conflicts and court cases in Florida and elsewhere in the United States. Therefore, they borrowed carefully from laws drafted elsewhere, including New South Wales in Australia and British Columbia in Canada, ensuring that management association bylaws, accounting, and shared maintenance provisions were carefully specified in the initial law (Fanjoy 1968, 184; Risk 1968, 24). Another feature of the Ontario case is that, while legislative revision has been infrequent compared with that in Florida, it has been careful and thorough, and large-scale legislative reform commissions have been established to research and draft legal revision in 1978, 1998, 2012, and 2017. So the processes of legal reform are different in various jurisdictions, as are institutional capacity and political dynamics.

Finally, even as condominium law changes and evolves over time, in almost all cases such changes are applied only to future projects, leaving existing properties unaffected. The Ministry of Justice in Japan and the Florida legislature explicitly refused to make new rules retroactive, arguing that doing so would be an encroachment on property rights and existing private contracts freely entered into. This means that condominium property created at different times, under different legal regimes, has different specified ownership characteristics and rights. And those differences are important, as witnessed by the high levels of court cases and conflicts in many jurisdictions. The rules for decision-making, awarding contracts, disclosure of contracts and finances, fees for maintenance, and special assessments for renovations can have a huge impact on the affordability of condominium units and their resale value. They profoundly affect the very meaning and nature of the property ownership itself.

The point here is that the systems that structure the production of new property change over time and have followed different evolutionary pathways in different jurisdictions. Although enacted within a short period and following apparently similar approaches, the legal frameworks of condominium in Japan, Florida, and Ontario started out different because of the underlying legal systems they were built on, and they evolved differently because of different political dynamics and processes of change. They are clearly not unchanging, but are evolving along different pathways. Condominium owners in the three jurisdictions face very different challenges and obstacles in each. And their products, in the form of new properties and bundles of property rights, are often enduring and can have significant variation among the basic forms of property that are produced. The regulatory and legal systems designed to control the creation of new property in urban areas, particularly in the suburbs, should therefore be central to the comparative study of cities and suburbs.

Conclusions

If we acknowledge that new and different forms of property are possible, as indicated by the invention and proliferation of condominium property, then the institutions designed to regulate the creation of new property in urban areas, particularly in the suburbs, become central to the comparative study of cities and suburbs. I suggest that to understand suburban development it is necessary to understand how property ownership is being transformed into a highly differentiated set of products that generate distinct political dynamics and path-dependent trajectories in different cities and under different legal regimes. I suggest that the rules structuring the creation of suburban property and the sets of property created are both geographically and temporally differentiated and produce path-dependent patterns of property rights attached to specific properties and sets of properties. The underlying logics of different sorts of property production, governance, servicing, and management are fundamental for understanding processes of suburbanization.

Processes of land subdivision and the creation of new property rights and ownership forms and patterns are thus a fundamental and generative characteristic of suburbanization. This perspective also underlines the central role of states' planning and legal systems in the definition and creation of property rights in cities.

In particular, privately governed housing developments with shared ownership of common facilities have become a major component of suburban development around the world, and they create new forms of suburban property ownership that have quite different political, property, infrastructural, and social logics than traditional freehold and leasehold forms. These changes have profound implications for people's relationships to place, homeownership, community, municipal governance, and local democracy, particularly as this new form of ownership has become more prevalent. These new ownership forms are also exceptionally path dependent, as they produce powerful incentives to block change that might undermine resale value.

The analysis also points to the centrality of systems that regulate the conversion of land into property in processes of suburbanization as a focus of comparative research on global suburbanisms. Understanding the regulatory systems that structure the creation of new property, and the variety of new property landscapes created is particularly important if we wish to compare patterns of development in different cities and countries, as contingent moments of legislative development and revision create increasingly varied legal regimes of property in different countries, which themselves are highly path dependent.

Comparing the institutions that regulate and shape the production of new property and property rights would answer the call by Robinson (2006) and others for new approaches to comparative urban research that are not premised on comparing the most similar cities in the global north, but are able to include a wider range of urbanisms into their comparative frame. By comparing particular governance institutions and their products, instead of the much more complex task of comparing whole planning systems or urban areas, it becomes possible to include a much wider range of cases. Comparison of the systems and institutions designed to regulate the conversion of undeveloped land into urban property and comparison of the resulting landscapes of property rights could therefore become a primary site of comparative urban research. Peri-urban and suburban areas are the primary location of such production of new property. A robust comparative analysis of the systems regulating the production of new property would also enable a mapping of the new geographies of property that are being produced by processes of suburban development surrounding cities around the world.

The meaning of land and land ownership is fluid, contested, and socially constructed. It is also profoundly shaped by legal systems of property and those evolve over time and place. The main suggestion here is that processes of suburban growth always involve establishing new sets of property rights and relationships. Yet the legal and institutional setting and requirements structuring such transformations show great variation in different places and evolve over time. Further, the kinds of property produced vary greatly, as the radical innovation of condominium makes clear. An understanding of suburbs as collections of property with variously defined property rights, in which the mix and configuration of the "bundles of rights" embedded in particular properties evolve over time encourages a new approach to the study of suburbanization, combining a genealogy of property law and a geography of property rights. The landscapes of property created during processes of suburbanization could thus be rendered visible.

REFERENCES

Adams, D. 1994. *Urban Planning and the Development Process*. London: UCL Press.

Adams, D., and S. Tiesdell. 2010. "Planners as Market Actors: Rethinking State-Market Relations in Land and Property." *Planning Theory & Practice* 11 (2): 187–207. https://doi.org/10.1080/14649351003759631.

Angel, S. 2012. *Planet of Cities*. Cambridge, MA: Lincoln Institute of Land Policy.

Blakely, E.J., and M.G. Snyder. 1997. *Fortress America: Gated Communities in the United States*. Washington, DC: Brookings Institution Press.

Blandy, S. 2006. "Gated Communities in England: Historical Perspectives and Current Developments." *GeoJournal* 66 (1-2): 15–26. https://doi.org/10.1007/s10708-006-9013-4.

Blomley, N.K. 1994. *Law, Space, and the Geographies of Power*. New York: Guilford Press.

Blomley, N.K. 1998. "Landscapes of Property." *Law & Society Review* 32 (3): 567–612. https://doi.org/10.2307/827757.

Blomley, N.K. 2004. *Unsettling the City: Urban Land and the Politics of Property*. New York: Routledge.

Blomley, N.K., D. Delaney, and R.T. Ford. 2001. *The Legal Geographies Reader: Law, Power, and Space*. Oxford: Blackwell.

Booth, P. 1996. *Controlling Development: Certainty, Discretion in Europe, the USA and Hong Kong*. London: UCL Press.

Buitelaar, E., A. Lagendijk, and W. Jacobs. 2007. "A Theory of Institutional Change: Illustrated by Dutch City-Provinces and Dutch Land Policy." *Environment & Planning A* 39 (4): 891–908. https://doi.org/10.1068/a38191.

Caldeira, T. 1996. "Fortified Enclaves: The New Urban Segregation." *Public Culture* 8 (2): 303–28. https://doi.org/10.1215/08992363-8-2-303.

Community Associations Institute. 2012. *Industry Data, National Statistics*. Accessed 8 October 2012, from http://www.caionline.org/info/research/Pages/default.aspx.

Cox, K.R. 2002. *Political Geography: Territory, State, and Society*. Oxford: Blackwell. https://doi.org/10.1002/9780470693629.

Coy, M. 2006. "Gated Communities and Urban Fragmentation in Latin America: The Brazilian Experience." *GeoJournal* 66 (1-2): 121–32. https://doi.org/10.1007/s10708-006-9011-6.

Coy, M., and M. Pöhler. 2002. "Gated Communities in Latin American Megacities: Case Studies in Brazil and Argentina." *Environment & Planning B* 29 (3): 355–70. https://doi.org/10.1068/b2772.

Darin-Drabkin, H. 1977. *Land Policy and Urban Growth*. Oxford: Pergamon Press.

Davis, M. 1990. *City of Quartz: Excavating the Future in Los Angeles*. London: Verso.

Dear, M. 1992. "Understanding and Overcoming the NIMBY Syndrome." *Journal of the American Planning Association* 58 (3): 288–300. https://doi.org/10.1080/01944369208975808.

DeGrove, J.M., and D.A. Miness. 1992. *The New Frontier for Land Policy: Planning and Growth Management in the States*. Cambridge, MA: Lincoln Institute of Land Policy.

Dredge, D., and E. Coiacetto. 2011. "Strata Title: Towards a Research Agenda for Informed Planning Practice." *Planning Practice and Research* 26 (4): 417–33. https://doi.org/10.1080/02697459.2011.582383.

Fainstein, S.S. 1994. *The City Builders: Property, Politics, and Planning in London and New York*. Oxford: Blackwell.

Fanjoy, E.O. 1968. "The Condominium Act (Ontario)." *Chitty's Law Journal* 16 (6): 181–5.

Fennell, L.A. 2009. *The Unbounded Home: Property Values beyond Property Lines*. New Haven: Yale University Press. https://doi.org/10.12987/yale/9780300122442.001.0001.

Fischel, W.A. 2001. *The Homevoter Hypothesis: How Home Values Influence Local Government Taxation, School Finance, and Land-Use Policies*. Cambridge, MA: Harvard University Press.

Goldsmith, M. 1995. "Autonomy and City Limits." In *Theories of Urban Politics*, ed. D. Judge, G. Stoker, and H. Wolman, 228–52. London: Sage.

Graham, S., and S. Marvin. 2001. *Splintering Urbanism: Networked Infrastructures, Technological Mobilities and the Urban Condition*. London: Routledge. https://doi.org/10.4324/9780203452202.

Harvey, D. 1982. *The Limits to Capital*. Chicago: University of Chicago Press.

Hayward, C.R. 2009. "Urban Space and American Political Development: Identity, Interest, Action." In *The City in American Political Development*, ed. R. Dilworth, 141–53. New York: Routledge.

Healey, P. 1992a. "Development Plans and Markets." *Planning Practice and Research* 7 (2): 13–20. https://doi.org/10.1080/02697459208722842.

Healey, P. 1992b. *Rebuilding the City: Property-Led Urban Regeneration*. London: E. & FN Spon.

Holder, J., and C. Harrison, eds. 2003. *Law and Geography*. Oxford: Oxford University Press. https://doi.org/10.1093/acprof:oso/9780199260744.001.0001.

Jacobs, H.M., and K. Paulsen. 2009. "Property Rights: The Neglected Theme of 20th-Century American Planning." *Journal of the American Planning Association* 75 (2): 134–43. https://doi.org/10.1080/01944360802619721.

Jürgens, U., and M. Gnad. 2002. "Gated Communities in South Africa – Experiences from Johannesburg." *Environment & Planning B* 29 (3): 337–53. https://doi.org/10.1068/b2756.

Kirby, A. 2008. "The Production of Private Space and Its Implications for Urban Social Relations." *Political Geography* 27 (1): 74–95. https://doi.org/10.1016/j.polgeo.2007.06.010.

Kohn, M. 2004. *Brave New Neighborhoods: The Privatization of Public Space*. New York: Routledge.

Kostof, S. 1991. *The City Shaped: Urban Patterns and Meanings through History*. Boston: Little Brown.

Krueckeberg, D.A. 1995. "The Difficult Character of Property: To Whom Do Things Belong?" *Journal of the American Planning Association* 61 (3): 301–9. https://doi.org/10.1080/01944369508975644.

Lasner, M.G. 2012. *High Life: Condo Living in The Suburban Century*. New Haven: Yale University Press.

Lee, S., and C. Webster. 2006. "Enclosure of the Urban Commons." *GeoJournal* 66 (1-2): 27–42. https://doi.org/10.1007/s10708-006-9014-3.

Low, S.M. 2003. *Behind the Gates: Life, Security, and the Pursuit of Happiness in Fortress America*. New York: Routledge.
Low, S.M., and N. Smith. 2006. *The Politics of Public Space*. New York: Routledge.
McAuslan, P. 1987. "Land Policy: A Framework for Analysis and Action." *Journal of African Law* 31 (1-2): 185–206. https://doi.org/10.1017/S0021855300009323.
McCaughan, R. 1964–5. "The Florida Condominium Act Applied." *University of Florida Law Review* 17: 1–55.
McFarlane, C. 2010. "The Comparative City: Knowledge, Learning, Urbanism." *International Journal of Urban and Regional Research* 34 (4): 725–42. https://doi.org/10.1111/j.1468-2427.2010.00917.x.
McKenzie, E. 1994. *Privatopia: Homeowner Associations and the Rise of Residential Private Government*. New Haven: Yale University Press.
McKenzie, E. 2003. "Common-Interest Housing in the Communities of Tomorrow." *Housing Policy Debate* 14 (1-2): 203–34. https://doi.org/10.1080/10511482.2003.9521473.
Mitchell, D. 2003. *The Right to The City: Social Justice and the Fight for Public Space*. New York: Guilford Press.
Needham, B. 2006. *Planning Law and Economics: An Investigation of the Rules We Make for Using Land*. London: Routledge.
Nelson, R.H. 2005. *Private Neighborhoods and the Transformation of Local Government*. Washington, DC: Urban Institute Press.
Poliakoff, G.A. 1992. "The Florida Condominium Act." *Nova Law Review* 16 (1): 471–513.
Risk, R.C.B. 1968. "Condominiums and Canada." *University of Toronto Law Journal* 18 (1): 1–72. https://doi.org/10.2307/825167.
Robinson, J. 2006. *Ordinary Cities: Between Modernity and Development*. Abingdon, UK: Routledge.
Robinson, J. 2011. "Cities in a World of Cities: The Comparative Gesture." *International Journal of Urban and Regional Research* 35 (1): 1–23. https://doi.org/10.1111/j.1468-2427.2010.00982.x.
Rolnik, R. 2013. "Late Neoliberalism: The Financialization of Homeownership and Housing Rights." *International Journal of Urban and Regional Research* 37 (3): 1058–66. https://doi.org/10.1111/1468-2427.12062.
Salet, W.G.M. 2002. "Evolving Institutions: An International Exploration into Planning and Law." *Journal of Planning Education and Research* 22 (1): 26–35. https://doi.org/10.1177/0739456X0202200103.
Sennett, R. 1977. *The Fall of Public Man*. Cambridge: Cambridge University Press.

Sies, M.C. 1997. "Paradise Retained: An Analysis of Persistence in Planned, Exclusive Suburbs, 1880–1980." *Planning Perspectives* 12 (2): 165–91. https://doi.org/10.1080/026654397364717.

Silver, C. 2008. *Planning the Megacity: Jakarta in the Twentieth Century*. London: Routledge.

Sorensen, A. 2015. "Taking Path Dependence Seriously: An Historical Institutionalist Research Agenda in Planning History." *Planning Perspectives* 30 (1): 17–38. https://doi.org/10.1080/02665433.2013.874299.

Sorkin, M. 2008. *Indefensible Space: The Architecture of the National Insecurity State*. New York: Routledge, Taylor and Francis.

Taylor, W.M. 2006. *The Geography of Law: Landscape, Identity and Regulation*. Oxford: Hart.

UNDESA. 2012. *World Urbanization Prospects: The 2012 Revision*. New York: United Nations Department of Economic and Social Affairs.

Upham, F.K. 1987. *Law and Social Change in Postwar Japan*. Cambridge, MA: Harvard University Press.

Valverde, M. 2012. *Everyday Law on the Street: City Governance in an Age of Diversity*. Chicago: University of Chicago Press. https://doi.org/10.7208/chicago/9780226921914.001.0001.

Walsh, E., R. Warland, and D.C. Smith. 1993. "Backyards, NIMBYs, and Incinerator Sitings: Implications for Social Movement Theory." *Social Problems* 40 (1): 25–38. https://doi.org/10.2307/3097024.

Warner, M. 2011. "Club Goods and Local Government: Questions for Planners." *Journal of the American Planning Association* 77 (2): 155–66. https://doi.org/10.1080/01944363.2011.567898.

Weber, R. 2010. "Selling City Futures: The Financialization of Urban Redevelopment Policy." *Economic Geography* 86 (3): 251–74. https://doi.org/10.1111/j.1944-8287.2010.01077.x.

Webster, C. 2001. "Gated Cities of Tomorrow." *Town Planning Review* 72 (2): 149–70.

Webster, C. 2002. "Property Rights and the Public Realm: Gates, Green Belts, and Gemeinshaft." *Environment and Planning. B, Planning & Design* 29 (3): 397–412. https://doi.org/10.1068/b2755r.

Webster, C., and L.W.C. Lai. 2003. *Property Rights, Planning and Markets: Managing Spontaneous Cities*. Cheltenham, UK: Edward Elgar.

Webster, C., G. Glasze, and K. Frantz. 2002. "The Global Spread of Gated Communities." *Environment & Planning B* 29 (3): 315–20. https://doi.org/10.1068/b12926.

Weir, M. 1995. "Poverty, Social Rights, and the Politics of Place in the United States." In *European Social Policy: Between Fragmentation and Integration*, ed. S. Leibfried and P. Pierson, 294–336. Washington, DC: Brookings Institution.

Weiss, M.A. 1987. *The Rise of the Community Builders: The American Real Estate Industry and Urban Land Planning*. New York: Columbia University Press.

West, M.D. 2005. *Law in Everyday Japan: Sex, Sumo, Suicide, and Statutes*. Chicago: University of Chicago Press. https://doi.org/10.7208/chicago/9780226894096.001.0001.

Williamson, T. 2010. *Sprawl, Justice, and Citizenship: The Civic Costs of the American Way of Life*. Oxford: Oxford University Press. https://doi.org/10.1093/acprof:oso/9780195369434.001.0001.

chapter four

Planetary Urbanization in Zurich: A Territorial Approach for Urban Analysis

CHRISTIAN SCHMID

Urbanization has shown remarkable changes in recent years. Urban areas are unravelling and expanding far into former peripheries, and new multi-scalar patterns of uneven socio-spatial development are emerging. New centralities are developing in once peripheral places, and new urban configurations are constantly produced and reproduced as the result of the proliferation of urban networks and lifestyles far beyond the confines and catchment areas of urban regions. The classic image of the city as an urban core surrounded by less dense and less central (sub) urban areas followed by a more or less rural hinterland has lost much of its empirical evidence and has thus to be radically re-conceptualized.

These changes make it necessary to rethink categories such as "urban," "suburban," and "rural" in order to adequately understand current trends of urban development. They demand a richer and more appropriate urban vocabulary, which would take us beyond the classic dichotomies of rural-urban, urban-suburban, and centre-periphery. Instead of trying to fit all emerging new configurations into the strait-jacket of well-established categories, we could develop a much more differentiated vocabulary of urbanization. Drawing on the basic concept of "planetary urbanization" (Brenner and Schmid 2014, 2015), this chapter traces current urbanization processes that can be observed in the Zurich region and proposes new urban typologies in order to develop a relational and dynamic understanding of urban territories that are constantly transformed and reshaped.

The Zurich region has experienced astonishing urban transformations in the last decades. While its economy was still strongly marked by manufacturing industries until the early 1970s, the crisis of the Fordist-Keynesian development model brought a radical socio-economic change.

During the 1980s Zurich developed into a small but highly connected type of global city, specializing in a narrow sector of financial industries and global headquarter functions. This change was accompanied by an equally fundamental urban transformation in turn accelerated by the microelectronic revolution and the resulting introduction of new communication technologies and the massive extension of transport infrastructure on all scales. The relatively compact agglomeration of Zurich extended its reach, stretching out further and further into neighbouring cantons, incorporating a range of smaller towns, villages, and former rural areas into a complex, multi-scalar, and polycentric metropolitan region.

While large parts of its inner-city neighbourhoods have been turned into privileged and gentrified urban spaces in the last few years and are functioning as business, entertainment, and consumption areas for the whole region, new centralities have emerged in the former urban periphery, especially in the northern area of Zurich adjacent to the airport. Already, two decades ago, the term "FlexSpace" (Lehrer 1994) was formulated in order to characterize these new urban configurations, which were marked by increased connectivity, high variability of urban structures, and high flexibility of land use, resulting in a surprising mixture of diverse urban functions, and generating a sometimes bewildering urban patchwork that was neither urban nor suburban in character. Since then, many other areas in the former urban periphery of Zurich have developed into some version of "FlexSpace," up to the point that highly specialized global city functions are now found even in once-remote places adjacent to farmhouses and meadows.

Recent rounds of urbanization have once again subjected these places to important transformations that profoundly challenge their interpretation as urban peripheries. Today, a new phase of urban development can be discerned that is marked by a process that we might call "urban intensification" (Nüssli and Schmid 2016). As the demand for "urban" living, working, and leisure strongly increased in the last decade, and as the inner-city neighbourhoods and even adjacent areas are already heavily "upgraded" and gentrified, "urban spaces" are now also sought after in the former urban periphery. Accordingly, local and regional authorities, planners, architects, and developers are searching for new ways to construct urban landscapes that can provide the highly valued features of "urban life" usually found in inner-city areas, such as diversity, density, access to public transport, public spaces, parks, and "urban architecture."

Today, the Zurich region thus is marked, like many other metropolitan regions, by its polycentrality, high levels of complexity, and strong internal differentiation. Different urbanization processes overlap and intermingle, generating a wide variety of urban configurations. Furthermore, this region cannot be understood without analysing the rapidly urbanizing landscapes located beyond its commuter zones and catchment areas. This chapter first presents a short discussion of the suburban question from the perspective of planetary urbanization. Second, it analyses the urbanization of Switzerland in order to explain the context of the patterns and pathways of urban development of Zurich. Finally, it presents an analysis of two urban configurations in the Zurich region to illustrate the wide variety of urban processes shaping contemporary urban territories.

Planetary Urbanization

In the last few years, the "suburban question" (Keil 2013b) has found new prominence in debates on urban development. This renewed interest in seemingly ordinary and unspectacular urban territories could be interpreted as a reaction to recent metanarratives of the contemporary global urban condition that have sometimes even taken the form of an urban triumphalism (see Brenner and Schmid 2015). However, as many observers confirm, the "suburban" is a highly heterogeneous category that encompasses a wide range of different constellations (see Keil 2013a). In most cases the "suburb" is only vaguely defined as located at the "periphery," or as a socio-spatial configuration that is outside the "urban." Additional criteria are sometimes included, such as lower density of the built environment and newness of the development (Harris 2010). But such definitions embrace an extremely wide range of urban configurations, whose main common features are that they are located outside the central areas of cities, and that they are "less" dense and relatively new. These criteria apply to so many different urban configurations and include such a wide variety of urban forms and processes that it is indeed questionable as to what analytical value we might find in such a generalizing concept.

Furthermore, this putatively "suburban field" has seen spectacular transformations in recent years, with the production of new centralities in once peripheral locations, various forms of densification, increasing diversity of land uses, growing heterogeneity and differentiation of social and ethnic composition, the installation of new government arrangements, and many more (see, e.g., Hamel and Keil 2015; Keil and Addie 2015; Phelps, Wood, and Valler 2010). Concepts such as "post-suburbanization" (Fishman 1991; Phelps and Wu 2011) or "in-between cities" (Sieverts 2000; Young et al. 2011) try to express some of these new developments but so far have not succeeded in developing a concise new conceptualization of the suburban field.

At the same time, so-called urban areas have also changed dramatically. The massive expansion of central business districts and the strong urban transformation of inner-city neighbourhoods driven by large-scale reinvestments, urban regeneration projects, and gentrification processes, often accompanied by the massive construction of condominium towers and the forced privatization of public housing, have led to the commodification, banalization, and mainstreaming of many former central areas, processes that can be summarized, beneath others, by the term "new metropolitan mainstream" (Schmid and Weiss 2004; Schmid 2011). As a consequence, many inner-city areas are today not the heterogeneous, vibrant, and restless neighbourhoods full of social interactions so often referred to in popular images and in scientific accounts of the "urban." Rather, they are developing towards pacified, antiseptic, controlled, and commodified spaces with precisely those characteristics that are commonly referred to as "suburban," such as uniformity of social composition and spatial form, monotony of daily life, and sterile and highly policed public spaces.

These processes have also profoundly transformed the familiar form of agglomerations and urban regions, which many are regarding as the main units of urban analysis (e.g., Storper and Scott 2016). An urban region could be loosely defined as the area of influence of an urban centre or as the area of active interactions between a centre and its surroundings, whereby its outer boundary could be determined by a significant drop in rate of flows and movements (Forman 2008, 6). This is usually represented with the well-known spatial image of a metropolitan node surrounded by its regional hinterland, forming the basic "unit" of a globalized economy more broadly constituted by an interdependent global network of metropolises (Scott 1998). However, the zones of influence of many urban regions have extended further in the last decades, stretching far out into formerly peripheral areas and profoundly transforming them. At the same time urban regions have become polycentric and their catchment areas have begun to overlap, which accelerates the development of multi-scalar urban realities. Throughout the history of urban studies, then, new terms have been constantly created in order to grasp these new emerging urban "units," from mega-city regions (Hall and Pain 2006) to "urban megaregions" (Harrison and Hoyler 2015) and even "urban galaxies" (Soja and Kanai 2006). With his concepts of "postmetropolis," and later of "extended regional urbanization," Soja (2000, 2010) tried to grasp precisely these kinds of polymorphic and "unbounded" urban areas and to express the frequent observation that the socio-spatial effects of urbanization "do not just decline with distance from the center to some outer boundary" (2010, 686). Processes such as the dispersion of once central functions and the creation of new centralities, the disintegration of the hinterlands, the densification of inter-metropolitan networks, and the emergence of new production zones outside the metropolitan cores, are further destabilizing the idea that it is possible to clearly identify and demarcate bounded and contingent urban areas. The urban fabric now extends further into adjacent territories than the concepts of "hinterland" or "urban region" allow for. We can also observe a profound socio-economic and environmental transformation of "rural" areas as well as the operationalization of erstwhile "wilderness" spaces and many more manifestations of extended urbanization (Brenner 2014; Brenner and Schmid 2015). Seen from this perspective, concepts such as "urban regions" as well as "suburban areas" appear in a new light: we cannot understand the urban region anymore as a neatly demarcated and bounded urban space. There is no clear border that could define such a region and draw a dividing line or zone between the "inside" (the urban area) and the "outside" (the seemingly non-urban realm). In a similar and related way the classic dichotomy between the urban and the suburban is put into question: Is it still possible to demarcate the urban and the suburban? Could the "suburban" still be considered fundamentally divergent from the "urban," or at least conceived of as a distinctive field that is marked by decisively different conditions?

Similar questions arise if we look at "suburban" areas from the "outside": there seems to be no end to the process of urban expansion – the urban outskirts are stretching far out into "rural" areas and transforming even wilderness zones – processes that are reflected in terms such as peri-urbanization, exurbanization, and *desakota* (see, e.g., Aydalot and Garnier 1985; McGee 1991; Taylor and Hurley 2016). Thus, the classic triad of urban theory, the urban – the suburban – the rural, is no longer applicable to many of today's urbanized areas. In order to reconfigure the conceptual formulations with which to grasp the extended landscapes of urbanization, we have to admit that the horizon of the analysis of urbanization today is the entire planet.

These kinds of urban transformation motivated the development of the concept of "planetary urbanization," which profoundly questions inherited understandings of the urban. Referring to Lefebvre's (2003) hypothesis of the complete urbanization of society, the basic idea behind this concept is that every part of the planet is today potentially confronted with, and marked by, urbanization processes in one way or another (Brenner and Schmid 2015). While there is no question that processes of concentration and thus the "power of agglomeration" still play a major role, urbanization also always entails processes of extension. Thus, planetary urbanization means a fundamental change in perspective: we should no longer focus on the development of a single "unit," a city or an urban region, but on the urbanized territory. Instead of drawing new boundaries and trying to define new types of urban units, it would be much more appropriate to try to understand different types of urbanized (and urbanizing) territories and to draw a much more nuanced picture of the differences that are produced in an urbanized world. These reflections follow the territorial approach of urban analysis developed in the framework of ETH Studio Basel in the last years (see Diener et al. 2016; Schmid 2015, 2016).

The Complete Urbanization of Switzerland

Following the approach sketched above, the challenge is to understand urbanization as a multidimensional and multi-scalar process and to analyse the urbanization of a territory instead of focusing on specific bounded areas like "cities" or "urban regions." In such an analysis we have to identify the patterns and pathways of urbanization; the articulation of centralities and peripheral areas through diverse regional, national, and globalizing scales; the specific territorial regulations; and the social, cultural, and economic differences unfolding in everyday life. The urban development of Switzerland can illustrate this approach and serves as an example for the analysis of planetary urbanization. While there is still no comprehensive account of the urbanization process in Switzerland, there are numerous important articles, books, and partial analyses upon which this account has been based (see Bassand, Joye, and Schuler 1988; Diener et al. 2006; Eisinger 2004; Eisinger and Schneider 2003; Hermann and Leuthold 2003; Jaggy and Racine 2007;

Racine and Raffestin 1990; Schmid 1989, 2012, 2014; Schuler et al. 2006). The main settlement area of Switzerland is the Swiss Plateau, with its gentle wooded hills and wide valleys, which runs from the Alps in the southeast to the foot of the Jura Mountains in the northwest. Since the industrial revolution the urbanization of this area has been determined by two basic conditions: the deep-rooted federalist political system of Switzerland and its strong communal autonomy on the one hand and the long-standing history of decentralized industrialization on the other. These two closely related factors have resulted in a specific pattern of urbanization that is marked by a dense and tightly woven urban fabric with relatively small urban centres.

Even today, the commune forms the central territorial unit regulating urban development; it not only is responsible for the zoning laws but also has the power to stipulate the tax rates, meaning that the extent, pace, and shape of urbanization are controlled mainly by the communes. These local political entities usually try to strongly defend their local autonomy against national and cantonal territorial regulations, and they are also often hesitant to cooperate with the neighbouring communes. One result of this strongly localized land-use regulation is that communal boundaries are deeply inscribed into the territory and are clearly visible on maps showing the urban footprint (Schmid 2006b).

As in other parts of Western Europe, industrialization in Switzerland started not merely in the towns but in those parts of the countryside where both labour and water power were available. The first factories were strung like pearls on a necklace along the watercourses of the Swiss Plateau and the Alpine foothills. Further industrial development led to the formation of three core industrial regions: (1) the Jurassic Arc and Geneva with its watch industry; (2) the region around Basel, with its textile industry that evolved in the twentieth century by way of the dye industry's shift into the chemical and later the pharmaceutical industry and the life sciences; (3) northeastern Switzerland, by far the largest industrial zone of Switzerland, based on the textile industry that evolved into the machine industry. Zurich, which at the end of the eighteenth century was a small town with around 10,000 inhabitants, became the very centre of this region and developed into the largest city in Switzerland as early as the end of the nineteenth century, based on its strong machine industry, its evolving banking sector, and its central position in the newly developed system of railway lines (see Bärtschi 1983; Walter 1994).

Despite some relatively modest forms of urban concentration, during the late nineteenth and early twentieth centuries this basically decentralized form of urbanization was accentuated and gave way to an early form of urban sprawl: extending from the densely woven pattern of existing towns and villages, the urbanized areas grew through radial expansion of the settlement areas in each of the communes. As a result, industrialized urban zones emerged in eastern Switzerland and along the southern base of the Jura Mountains that consolidated into an almost continuous urbanized belt from Lake Constance to Biel, whereas in well-located Alpine regions the first touristic zones emerged. Just a few towns developed into larger cities, and only Zurich, Basel, Geneva, and Berne counted more than 100,000 residents in 1930.

This decentralized urbanization process was reinforced through a distinctive type of Helvetic Fordism after World War II. It was not primarily based on mass production, because Swiss industry (especially the watch and machine industries) increasingly specialized in labour-intense, high-quality, and technologically advanced consumer and investment goods produced by small to medium-sized companies with predominantly qualified workers. As a consequence, most companies expanded their production facilities at their existing locations, resulting in a reinforcement of the extensive urban growth pattern. Thus, Switzerland's decentralized industrial structure was largely preserved even in the Fordist-Keynesian development phase, and the belt-like urbanization in eastern Switzerland and along the southern base of the Jurassic Arc was consolidated and densified.

In addition, the Swiss version of a Keynesian welfare state implied a model of spatial development that was driven by a strategy of "equality in space," according to which all parts of this highly decentralized country should enjoy the same high standard of infrastructure and public services (Blanc and Luchsinger 1994; Marco et al. 1997). This strategy set in motion a comprehensive process of social modernization, which in turn led to a certain standardization of the territory, mainly advanced by the construction of a national highway network, which provided almost every small town with a direct link to urban life and subsequently became the main driver of urban development (Studer 1985). Thus, Helvetic Fordism resulted in a decentralized and scattered pattern of urbanization, strongly marked by the inherited territorial cell structure of the communes, which aspired to attract as many urban functions as possible to their small territories, residential zones for good taxpayers as well as industrial and commercial zones. As a consequence, the urbanization process was dominated by the radial extension of the tightly woven net of villages and towns, generating a small-scale pattern of urbanization at the Swiss Plateau and prompting a specifically Swiss form of urban sprawl (see Corboz 1988; Meili 1941; Schwick et al. 2012).

The mid-1970s brought a dramatic change in this decentralized urbanization pattern, when, as in almost all industrialized nations of the West, the Fordist-Keynesian model of development experienced a deep crisis. Switzerland has been strongly affected by this crisis, which resulted in massive economic restructuring and a radical reconfiguration of the industrial landscape. While less skilled jobs in the manufacturing sector were either eradicated or relocated, highly qualified economic activities, especially headquarter functions, strongly increased. As a result, a divided spatio-economic structure emerged (see Crevoisier et al. 2001). On one hand, nationally protected economic sectors, such as tourism, local crafts, and agriculture (which dominated the more peripheral regions of Switzerland) remained largely untouched by rationalization. Remnants of the manufacturing industry, including the high-tech sector, continued to develop in a geographically dispersed way (Dümmler et al. 2004). On the other hand, a "headquarter economy" evolved, specialized in organizing transnational production processes and controlling global capital flows (Hitz, Schmid, and Wolff 1994, 1995; Schmid 2006a). As a result, the main urban centres of Switzerland developed into highly specialized economic complexes: life sciences in Basel, specialized financial services in Geneva, and global finance in Zurich. Most notably, with ever new rounds of deregulation of national and global financial markets, the financial complex of Zurich developed into an important node of global financial circuits, specializing in the organization and control of global production and finance, and thus became an important driver of globalization. Despite its relatively small size, Zurich developed into a global city (Brenner and Keil 2006; Sassen 1991; Schmid 1998, 2006a).

This radical economic restructuring was accompanied by similarly far-reaching changes in the urbanization process. Alongside the persistence of the prevailing decentralized model of urbanization, a new dynamic of concentration of population and economic activities occurred, in which the metropolitan nodes of Switzerland developed into vast polycentric urban regions, often integrating several urban centres into one overarching catchment area, which resulted in overlapping networks of regional and global reach and in a fragmented and unevenly developed urban fabric. As a result of this process of metropolitanization, three main metropolitan regions can be identified today: the Région Lémanique along the northern shore of Lake Geneva, with Geneva and

Figure 4.1 Typical articulation of FlexSpace: Pfäffikon, Switzerland. Photo: Caroline Ting, 2017

Lausanne as its two poles; the tri-national and tri-polar region emerging in Switzerland, Germany, and France, whose poles are Basel, Mulhouse, and Freiburg; and the region of Zurich (see Ascher 1995; Bassand and Schuler 1985; Cunha and Schuler 2001; Glanzmann et al. 2006; Leresche, Joye, and Bassand 1995; Seifert et al. 2006; Thierstein et al. 2006; Veltz 1996). However, according to the strongly decentralized urban landscape of Switzerland, the interlinked, polycentric urban regions that emerged were not confined to the big metropolitan centres but also became a feature of medium-sized towns that are more tightly connected with each other, forming "networks of cities" (see Diener et al. 2006; Krautzig, Gunz, and Mueller Inderbitzin 2012).

In general, metropolitanization in Switzerland has involved the dispersion of centrality: numerous new centralities – shopping malls, outlet stores, multiplex movie theatres, recreational facilities, hotels, clusters of office blocks, and even corporate headquarters – mushroomed in the urban peripheries of the three large metropolitan regions, especially in the north of Zurich (see below) and in Lausanne Ouest (Coen and Lambelet 2011). The same kind of dispersion of centralities and of various urban fragments emerged also in the shapeless territories between the smaller cites of the Swiss Plateau, for instance, in the amorphous urban belt developing south of Aarau and Olten (Oswald, Baccini, and Michaeli 2003), and in the area around Mendrisio in the southern part of Switzerland. As a result, patchwork-like urban landscapes marked by strong territorial, social, and economic fragmentation emerged in many parts of Switzerland, riven by manifold cultural and political fault lines.

Under the influence of metropolitanization, then, the old pattern of relatively dense, but decentralized industrial belts and concentrically expanding but small agglomerations is today dissolving into a pattern of great complexity and heterogeneity. Viewed in its entirety, the urbanization process is losing its contours and is becoming increasingly directionless and indiscriminate (see Corboz 1988, 2000; Schmid 2012). Furthermore, the urbanization process is penetrating ever deeper into remote areas and Switzerland's remaining agricultural landscapes – a process that the term "peri-urbanization" only approximately describes (Aydalot and Garnier 1985; Garnier, 1985; Schuler, Perlik, and Pasche 2004). Small towns and villages are increasingly being interspersed with zones of detached houses, industry, and commerce, thus transforming what were once "rural spaces" into new urban landscapes characterized by their incorporation into the urban networks of metropolitan regions, the dominance of extended forms of urbanization, and the accompanying fundamental changes in everyday life (Diener et al. 2006; Gunz and Mueller Inderbitzin 2008). Indeed, the zones experiencing the fastest population growth in Switzerland today are not only the centres of the metropolitan regions, but also the outskirts stretching out to the hilly landscapes of the Prealps.

Urbanization is even penetrating the areas beyond these zones and also transforms the heartland of putatively "rural" Switzerland – the Alps. We could therefore speak of the urbanization of the Alps: as detailed analyses show, large parts of the entire Alps are today strongly influenced by the extents of large metropolitan regions located at the vicinity of the Alpine arc, such as Vienna, Munich, Milan, Turin, Zurich, Région Lémanique, Marseille, and Nice (Bätzing 2015; Perlik 2001, 2010). Smaller agglomerations are expanding inside the Alps, around important urban centres such as Grenoble, Aosta, Brig, and Innsbruck. Also, many Alpine resorts today are integrated into complex networks of tourist infrastructure that link various catchment areas across the watersheds in different valleys. At the same time, uneven development becomes more pressing: while those Alpine regions that are readily accessible and have attractive winter sports arenas with sophisticated tourist infrastructure have become major resorts, other parts of the Alps with inferior transportation links or those that are unlikely to attract mass tourists are being left behind; their social and economic resources become depleted and they are developing into "Alpine fallowlands" (Diener et al. 2006). A new pattern of uneven spatial development is evolving, with boom regions like Milan and Zurich in the direct vicinity of impoverished Alpine valleys.

As the case of Switzerland illustrates, a planetary approach to urbanization that starts with an analysis of the urbanization of the entire territory rather than with an examination of distinct, bounded settlement types allows us to better understand the wider context of urbanization, to identify a series of urbanized and urbanizing landscapes (some of them located far away from urban centres), and thus to define the framework for the analysis of the Zurich region. Placing the generic and generalizing concept of the "suburb" in the broader context of the urbanized territory of Switzerland, we can clearly see that a wide range of peripheral types of urbanization have evolved, not only in the outskirts of the metropolitan regions, but also between midsize centres in the Swiss Plateau and in the wider context of the once rural areas in Prealpine and Alpine regions.

The Implosion/Explosion of the Zurich Region

Departing from this background of urbanization in Switzerland, we can zoom in and analyse the differentiation of urbanized territories in the Zurich region. As the previous analysis has shown, this region has seen a dramatic socio-economic change from an industrial agglomeration towards a polycentric metropolitan zone in only two decades, from the mid-1970s to the mid-1990s. While most of the jobs in the manufacturing sector were either rationalized or relocated to other places in Switzerland or abroad, more than one-third of all jobs are now in the core sectors of the headquarter economy, such as the financial industry, global insurance, and producer services. Until the 1970s most of these economic activities were concentrated in a small area in the inner city of Zurich. With the upswing of the headquarter economy a kind of implosion-explosion of the region occurred, and these activities were scattered across a vast area in the urban periphery. This led to a complex process of urban restructuring, marked by processes of urban extension and intensification, the regionalization of the economy and the formation of a highly fragmented regional scale of urban governance.

As a result of this process of implosion-explosion (Lefebvre 2003), the emerging "Zurich region," like many other metropolitan regions, forms no coherent and neatly bounded unit that could be clearly defined and delimited. The term "metropolitan region" serves mainly as a metaphor for an urban territory that is polycentric, fuzzy, fragmented, and disputed. In order to define a regional "unit," quite different criteria and indicators could be selected, resulting in a great variety of different possible "regions" (see Schmid 2007). By analysing economic networks and the regionalization of the headquarter economy, an area that covers large parts of northern Switzerland has been identified as a "European Metropolis Region" with about 2 million inhabitants (Glanzmann et al. 2006). From the perspective of everyday life, however, the Zurich region has a different shape, which basically depends on centralities, catchment areas, and commuting patterns. The opening of a new metropolitan railway system (S-Bahn) in 1990 was decisive in this respect, because it became the new reference frame for everyday life by structuring daily spatial practices but also by creating a new regional identity. This region today counts about 1.5 million inhabitants.

In contrast to these processes of regionalization, the political and institutional structure of the Zurich region is still strongly fragmented. Owing to the impact of the inherited federalist structure outlined above, no relevant territorial organization has been created that could develop and implement an urban strategy for the entire region. In recent years a *Metropolitankonferenz* (conference for the metropolitan region) has been installed; it does not form an operative unit but rather serves mainly as an institutionalized framework for discussions between different cantons and municipalities (see Nüssli 2015). To a considerable extent, the canton of Zurich should be understood as a "regional" governmental unit with about 1.2 million inhabitants, steering a large portion of the whole metropolitan area. The canton is responsible, among other things, for public transport, regional infrastructure, and the coordination of urban planning, and it would have the political and administrative power to govern urbanization. However, for a long time, the canton did not fully deploy this potential. While the planning department of the canton developed sophisticated strategies for the coherent urban development of the Zurich region, they were never implemented, because of the strong power of the communes. In fact, individual communes have dominated the planning process for decades, which has resulted in marked fragmentation of land-use patterns and strong urban sprawl across the emerging Zurich region. It was only recently that the canton of Zurich was willing and able to launch and implement a more coherent project of territorial planning that restricts urban sprawl and promotes urban densification and urban qualities.

Partly as a result of weak regional planning and fragmented governance structures, the Zurich region is marked by strong socio-spatial and morphological differentiation. Based on the analysis of socio-economic characteristics, urban densities, land-use patterns, and centralities, a wide range of urban configurations can be identified (see Schmid 2018): (a) inner city areas, still well preserved and marked by the presence of strong centralities and high densities, at the same time subject to massive upgrading and gentrification processes; (b) old centralities in transformation: long-existing and independent midsize towns like Zug, Winterthur, and Baden that have become incorporated into the metropolitan region; (c) the areas along the northern and southern shores of the lake of Zurich with entrenched affluent residential areas; (d) relatively dense types of urban configurations characterized by strong polycentrality and heterogeneity, currently in the process of urban intensification; (e) former industrial villages forming belt-like commuter zones without strong centralities; (f) mixed agricultural and industrial zones that are forming "double peripheries," as they are oriented towards two or more urban centres; (g) strongly globalized former rural landscapes; and (h) a series of different urban configurations that still have a strong portion of agricultural surfaces (often analysed as "peri-urban").

According to the general definitions presented at the beginning of this chapter, most of these very different urban configurations would be labelled "suburban," only with the exception of (a) the inner-city areas and (b) the old centralities. A more differentiated analytical perspective, however, reveals the great complexity of these landscapes (see also ETH Wohnforum 2010). I selected two of these urban configurations for presentation here: (1) the airport region in Zurich North highlights the dramatic urban transformation of an old commuter zone at the periphery into a patchwork-like "FlexSpace," and then towards a much more coherent and dense urban landscape strongly marked by the process of urban intensification; (2) the second case study, Outer Schwyz, serves as an example of the spectacular transformation of a former rural zone at the periphery into a peculiar fragment of a global financial centre.

Zurich North: From the Urban Periphery to a Landscape of Urban Intensification

The airport region in Zurich North is a telling example of urban developments that profoundly unravelled former urban peripheries in the Zurich region (for more detailed analyses of Zurich North see, e.g., Architektengruppe Krokodil 2012; Campi, Bucher, and Zardini 2001; Hitz, Schmid, and Wolff 1994, 1995; Lampugnani et al., 2007; Nüssli and Schmid 2016; Salewski and Michaeli 2012; Schubarth and Auderset 2004; Thierstein et al. 2005). For a long time seen as an "ordinary" commuter zone at the periphery, it is today mutating into a completely new type of urban configuration. Since the 1960s three phases of urban development can be distinguished: (1) the phase of the production of the urban periphery; (2) the phase of the generation of a multilayered urban patchwork; and (3) the most recent phase of densification and urban intensification.

Until the middle of the twentieth century the area north of the city of Zurich was marked by the decentralized industrial and urban development of Switzerland analysed above; it could even serve as a typical example of the resulting pattern of industrialized villages and small towns embedded in a landscape of gentle hills with forests and agricultural land. In the 1960s this area became a kind of a spillover basin, where all sorts of uses that did not find a place in Zurich's inner-city areas could be relocated. The term "urban periphery" was appropriate to designate this vast area at the outskirts of the city, marked by a typical heterogeneous mix of peripheral urban elements, such as industrial and infrastructural areas, agricultural land, traditional village cores, old working- and middle-class neighbourhoods, and some upper-middle-class residential areas at southern hillsides (see Nüssli and Schmid 2016). New centralities were established with the construction of Zurich's airport in Kloten and a huge regional shopping mall in Wallisellen. As a result, a polycentric urban fabric emerged in this area with centres along the highways rather than along the railway lines.

This situation changed radically in the 1980s. With the rise of the headquarter economy Zurich North became an important location for new office buildings. As the further expansion of office spaces in Zurich's inner-city areas was strongly limited, many companies moved their back offices and even their headquarter functions to the urban periphery, further attracted to the area because of its proximity to the airport, affordable land prices, and considerably lower tax rates (a direct consequence of municipal autonomy) (see Schmid 2006a).

In contrast to the city of Zurich with its strong regulation of urban development, the political fragmentation that resulted from municipal autonomy led to a very different pattern of urbanization. Politically, the core area of Zurich North is divided into the northern neighbourhoods of the city of Zurich and four independent municipalities: Kloten (the location of the airport), Opfikon, Wallisellen, and Dübendorf. Depending on the definition, at least eight further municipalities could be seen as being part of Zurich North in a broader sense (Schubarth and Auderset 2004). Because the individual municipalities were competing with each other to attract new enterprises in order to optimize land uses and tax income, they barely coordinated planning policies, which resulted in a haphazard juxtaposition of various uses and functions. A new polycentric and excentric urban configuration emerged, marked by a complex patchwork of more or less disjointed urban fragments on the territory. The new centralities were usually located at the periphery of the municipalities, where large plots of land for new uses were still available, either on industrial brownfield areas or on agricultural land. Thus, an odd belt composed of new centralities and business districts meandering through the whole area came into existence, often overlapping with old industrial zones and derelict land, whereas the old village cores were relegated to a peripheral status. The result was an urban landscape with open land and wide perspectives, at the same time defined and divided by infrastructural lines, such as railways, motorways, and runways that cut into the existing urban fabric. In this way very different elements were placed in direct spatial vicinity to each other: agriculture, old village cores, new urban extensions, old settlements, and residential areas with detached houses. This led to a reorientation of the urban fabric, which is now characterized by a patchwork of isolated basins that are separated from each other through infrastructural lines, each of which is filled with unrelated urban fragments laying side by side, but generating hardly any form of interaction (Kretz and Christiaanse 2018).

Precisely in relation to this example of Zurich North, Ute Lehrer created her concept of "FlexSpace," highlighting the strong relationship of this form of urbanization with wider socioeconomic processes of flexibilization and deregulation, produced by the flexibility allowed by massive transport infrastructure and the complex interplay of juxtaposed but specialized and isolated urban fragments (Lehrer 1994, 2013). In a similar way, in order to grasp these emergent urban forms, other concepts were proposed such as "network city" (*Netzstadt*) (Oswald, Baccini, and Michaeli 2003), in order to highlight the strong dependence of such areas on urban networks instead of physical proximity; or "in-between city" (*Zwischenstadt*) (Sieverts 2000), in order to stress that these areas are located between already urbanized landscapes but not at the edge of one – as the famous term "edge city" indicates for similar developments in North America (Garreau 1991). Despite the obvious difference in scale, this urban configuration could also be compared in many respects to the concept of "Exopolis," as developed by Edward Soja (1992), using the example of Orange County, which is located at the edge of the metropolitan region of Los Angeles. However, these terms are barely used any longer, as they mainly described "new" urban forms at that time; after the further urban transformation of the areas, they lost their explanatory power. Because they conceptualized certain urban forms, these concepts reflect historical moments in a located process of urban development.

In an attempt to develop a more general and widely useful characterization of this type of urban development, in recent comparative research we coined the term "Mulapa" (multilayered patchwork urbanization; Wong et al. 2017) to designate the specific urban processes that have led to these urban forms. Mulapa must be understood as a confrontation of different logics of urbanization leading to the interference of multiple spatial orientations and multi-temporal dynamics, which result in the accumulation and juxtaposition of diverse and unrelated urban elements within the territory over a long period. Such areas are usually determined by strong socio-spatial fragmentation and a highly complex and polycentric urban pattern. They can be found in many different places (our cases were in Los Angeles, Tokyo, Hong Kong, and Paris). Unlike the term "suburb," Mulapa does not refer to a general tendency towards urbanization, but designates one possible pathway of urban development beneath many others that often can be observed even in the same urban region (as in the case of Zurich). Furthermore, as with all types of urban development, it is not stable but evolves further.

Thus, over the last few years, Zurich North has experienced a new round of urban transformation, and a new kind of urban development can be identified. These changes are the result of a double process. First, massive but uncoordinated urban development generated through the political fragmentation and the dominance of the local scale of urban

Figure 4.2 Urban intensification in Zurich North: Glattpark, Opfikon, Switzerland. Photo: Caroline Ting, 2010

planning resulted in such a dispersed urban fabric that it became almost impossible to connect the newly built office zones and settlements with public and private transport; any further development of the area was crucially dependent on a more coherent urban structure. Second, the rise of Zurich's headquarter economy created a tendency towards a much stronger demand for "urban" living, which could no longer be met by the existing housing stock in the inner-city neighbourhoods. Both aspects strongly motivated the construction of new "urban" neighbourhoods in the former fragmented and dispersed urban periphery. In a certain way, the "new metropolitan mainstream" had arrived in Zurich North (see Schmid and Weiss 2004).

In recent years an astonishing urban transformation could be observed. A main instrument of this change was the construction of a new tramline sought to repair the fragmented urban pattern and to allow better connectivity between the various centralities of Zurich North. But at the same time the tramline also produced a new orientation of the entire area and became the backbone of an "urban" revival. Further urban growth resulted in a much denser urban fabric, and at the same time there have been important efforts in many places and on various scales to generate new connections between isolated urban islands. Concurrently, new forms of cooperation were established: the extreme political fragmentation that so strongly affected urban development gave way to new urban governance arrangements, including much stronger cooperation and professionalization of planning and territorial regulation (for a detailed analysis of this process see Nüssli and Schmid 2016). In addition, there have been great ambitions to create new public spaces, an urban image, and even an urban atmosphere that have proved at least partially successful. Such changes indicate a distinct period in the urbanization of Zurich North that we call urban intensification (ibid). The result is an urban configuration that provides decent urban spaces for the headquarter economy and for middle-class housing, but also generates socio-spatial exclusion – housing prices in these newly built areas are lower than in privileged inner-city areas, but nevertheless remain unaffordable for low-income groups.

Outer Schwyz: From a Rural Zone to a Globalized Village

With the rise of Zurich's headquarter economy, one of the most astonishing new types of urban configuration to emerge in the Zurich region is an area called Ausserschwyz (Outer Schwyz), which is located directly adjacent to the affluent residential areas strung out along Lake Zurich. This case reflects well the ways in which globalizing economic processes intersect with local political constellations to generate distinctive urbanized territories.

The strong valuation of Lake Zurich as an amenity has produced a long series of affluent neighbourhoods forming a kind of a linear city, a "built arena" stretching out along both shores of the lake (see ETH Studio Basel 2010). At the southern shore, about 20 km behind the city centre of Zurich, this situation changes abruptly. After the border into the canton of Schwyz is crossed, the landscape turns into a bewildering combination of rural elements, meadows and cows, small refurbished villages, small-scale industry, and detached houses, as well as a regional shopping mall, a regional cultural centre, and a water fun park. The most astonishing aspect, however, is that this place has developed into a location for corporate headquarters and hedge funds and thus has become a satellite of Zurich's headquarter economy. At the same time it has also become a discreet and quiet location for wealthy and very wealthy residents from all over the world, such as tennis star Roger

Federer, who lived here for several years. A typical architectonic and urbanistic expression of this global tax haven is the specific form of a condominium house, a kind of a vertical stapling of detached houses built on the slope with beautiful views of the lake and the mountains, where each apartment has a large terrace, constructed in such a way as to shield any direct view from the street to the house.

This area has a double geographical orientation. On the one hand, it is directly linked to the centre of Zurich and forms a satellite of Zurich's headquarter economy. It is strongly dependent on all the assets and utilities offered by the metropolitan region and it can function only because of its integration into this region. On the other hand, it shows a strong cultural orientation towards the more traditional regions of Obersee and Schwyz, thus strengthening the polycentric character of the metropolitan region (for a detailed analysis see Kretz and Christiaanse 2018; Kretz and Küng 2016; Nüssli 2017).

In the 1960s this area was a quiet backwater of Zurich, a peripheral and relatively poor rural and semi-industrialized area outside the main zone of influence of the agglomeration of Zurich. A decisive change occurred when the motorway opened in 1968, which brought the first wave of peri-urbanization to the area, as described earlier for the many remote areas of Switzerland at this time. The second wave of urbanization started in 1984, when the canton of Schwyz as well as the communes of Outer Schwyz decided to introduce a policy of radical tax breaks, using their cantonal and communal autonomy to profit from the increased connectivity that provided fast access to all the amenities and utilities offered by the entire region of Zurich. In the following years companies as well as wealthy people relocated to this area. Enterprises of the financial industry followed, especially hedge funds, which generated their massive profits with only a few employees and thus were not dependent on the kind of infrastructure characteristic of more central locations.

Here, then, we see the emergence of a new urban form in Switzerland: the globalized village. As in many other "classic" agglomeration areas, an amorphous patchwork structure of incoherent urban fragments evolved, but with a distinctive social and economic composition. The rural qualities of the territory are still visible, social rules and structures of village life are still present in many respects, and the regulation of the territory itself bears strong traces of this rural past. Important aspects of village communal governance persist, and even some medieval elements survive: the most important landowners in

Figure 4.3 Globalized village: hillside in Wollerau, Switzerland.
Photo: Caroline Ting, 2017

the area are the monastery of Einsiedeln and the old land corporation of Pfäffikon, originally founded to organize communal peasant life. These two actors are today the main agents of urban development (for a detailed analysis of this very peculiar form of territorial regulation see Nüssli 2017).

The extreme governmental fragmentation at the communal level, with a wide array of different internal arrangements for decision-making, as well as the patchwork nature of individual and traditional land ownership, has facilitated an uncoordinated, and even shocking, juxtaposition of different kinds of land use. Entrenched traditional social and spatial structures are confronted with the diverse fragments of the highly globalized satellite of the headquarter economy. As a direct result of the extremely low tax rates, land prices have increased massively, and while some landowners have benefited alongside the land corporation and the monastery, it has become difficult for many entrenched residents to stay in the area. In fact, this could be understood as a specific form of "gentrification," producing a strange island of luxury. The same phenomenon is also affecting similar locations, especially the canton of Zug.

An appropriate understanding of this urban configuration must also include the ways in which the formerly "rural" landscape is now used: all sorts of urban fragments are scattered over the whole area, destroying any illusion of being in the countryside. Consequently, the green areas between the built-up areas must be reconfigured as an "urban green," occupied by varied uses, from regional tourism to jogging or promenading with the dog (Muri 2018).

Here, the question of the "urban" plays a very different role than it does in Zurich North. Importance is placed on qualities such as separation, seclusion, and confidentiality, but not on lively public space. Indeed, the logic of these developments is not intended to produce any "urban" qualities. The local residents want to preserve their rural values, the internationally oriented inhabitants are not interested in local life, and both groups want to keep tax rates as low as possible. The "urban" densities and lifestyles or the vibrant public spaces so highly valued in other parts of Zurich are not sought. This attitude became archetypically evident when the commune of Wollerau decided to replace its old village square. The classic centre of the village, with a church, a hotel, and a restaurant, where traditional events and festivities of village life such as weddings and funerals took place, was demolished and replaced with a roundabout, which gives "efficient" access to the nearby motorway, but at the same time inhibits any public activity (Kretz and Küng 2016, 132–3). This situation can be seen both as a symbol for the eradication of the old communal life in the village and as an expression of the refusal of urban life.

This strange combination of a degraded village life and the anonymity of the metropolis, or more specifically the exclusivity of elite residents and the external orientation and connectivity of businesses, has produced a distinctive form of urbanized territory. Thus, life in these globalized villages becomes a kind of neither/nor – neither traditional village life with its intrinsic qualities, nor urban life with its specific potentials. This area lacks many attributes of urban life, but at the same time it is clearly not "rural" anymore. Using a territorial approach to understanding (planetary) urbanization, it is possible to build a distinctive but also generalizable analysis of this area as one among many other constantly transforming configurations of the urban.

Conclusion: The Question of the Urban Periphery Revisited

Over the past decades, the urbanization of Switzerland has generated fundamental transformations in settlement patterns and urban landscapes, and hence in living conditions as well. In that process what has been seen as the "suburb," the "urban periphery," or the "agglomeration zone" has profoundly changed. These once relatively uniform zones, which were planned and structured mainly by individual communes and covered large parts of the Swiss Plateau surrounding the large and medium-sized urban centres, have developed into highly complex, multi-scalar, differentiated, and sometimes even bewildering urbanized landscapes. Depending on the concrete situation, many different urban outcomes evolved. The examples presented here are just two out of many new urban configurations that can be identified in the urban region of Zurich and beyond.

Each of the two urban configurations analysed in this chapter shows a very characteristic trajectory, strongly influenced by its pre-existing urban fabric and its position in the regional pattern of centralities, as well as by specific local traditions, socio-economic conditions, and political constellations. Zurich North, having a strong industrial background and a relatively dense urban fabric, developed first into a complex multilayered patchwork of variegated urban elements. It then experienced a second phase of urban transformation, strongly shaped by various attempts from communes, cantonal agencies, planners, and developers, to create a "more urban" space serving the needs of the headquarter economy and the metropolitan middle classes. In this process the urban fabric became more tightly woven and internally connected, public transport was strongly improved (especially through the construction of a new tramline), and new parks and public spaces were created. This urban intensification was accompanied by a marked change in the social composition, as the newly built settlements are relatively expensive and therefore attract mainly middle classes, whereas lower-income groups living in the existing urban areas are gradually displaced towards more peripheral places. A similar model of urbanization is currently developing in the Limmattal west of Zurich, where a new tramline will soon be constructed. These two urban areas are thus considerably extending the densely settled "urban" core of the Zurich region towards the north and the west.

In contrast to these developments, Outer Schwyz, the other example presented here, is still strongly marked by the traditional social and political structures of a vanishing rural society that sometimes can be traced back even to medieval times. Here, too, a complex pattern of urbanization has emerged, leading to the uncoordinated juxtaposition of unrelated elements with a strong presence of the headquarter economy. Nevertheless, this area is very different from Zurich North because it is characterized by a different mix of urban elements, and it follows a very different logic of territorial regulation. The good connectivity to the centre of Zurich, the attractive landscape at the shore of the lake, and the autonomous regulation of tax regimes and land use offered the option of attracting global financial firms as well as internationally oriented wealthy people. The still entrenched rural identity of the local people and the desire for seclusion of the wealthy newcomers have led to a peculiar sociocultural constellation with a strong anti-urban bias. This urban change becomes clearly visible in the lack of public spaces, in the low density of the urban fabric, and in the fragmented urban structure. Despite the area's lack of urban intensity, it is nevertheless strongly confronted with urban problems like gentrification, traffic jams, and planning frictions. It marks another model of urban development that can be found in other privileged areas that are well connected, have favourable tax regimes, and nice landscapes, such as Zug or Obwalden, places located at quite a distance from Zurich but clearly still in the sphere of its economic influence.

As these examples show, there is not just one type of urban periphery developing around Zurich, but a great variety of urban configurations marked by different patterns and pathways of urbanization sometimes to be found far beyond any delimitation of the urban region. With a territorial approach inspired by the concept of planetary urbanization we can understand a metropolitan territory as being produced by a multitude of urbanization processes, resulting in an array of different urban configurations to form a sort of region. However, this region should be understood not as a bounded and clearly discernable or definable territorial unit, but rather as the temporary shape of the constantly evolving pattern of urbanization in the continuum of a multi-scalar urban landscape.

These metropolitan territories are defined not by one centre, but by a wide range of specialized and networked centralities. They thus constitute an assemblage of several interconnected and sometimes overlapping catchment areas, which have developed over the last decades in response to varied processes of globalizing economic activities and variously configured and fragmented processes of governance. Instead of studying single agglomerations or catchment areas, we must analyse the ensemble of these centralities and their mutually related peripheries in the context of a continuous but complex urban topography composed of different urban configurations.

Finally, it has become clear that the distinction between urban and suburban areas no longer proves useful. The dichotomy between these two categories not only is blurred, but becomes even misleading as an analytical tool. The horizon of the urban has changed: it is no longer possible to contrast the image of the urban with the rural background, or the suburban with the urban. As the Swiss case so eloquently demonstrates throughout its history of urbanization, various urbanized areas can be found across the entire territory. This makes it more difficult to identify and qualify the urban. Using the example of Zurich, this chapter has suggested an alternative, territorial approach to understand urbanization, to detect and decipher different kinds of urbanizing territories, and to develop a new vocabulary that helps us to discern the transforming and emerging forms of the urban.

ACKNOWLEDGMENTS

This article is a result of the research project, "Urban Potentials and Strategies in Metropolitan Territories at the Example of the Metropolitan Area Zurich," conducted in the framework of the national research program NRP 65, "New Urban Quality" (see Kretz and Küng 2016). I thank all my colleagues at the research project for their great cooperation, especially Rahel Nüssli, who worked with me on the case studies on Zurich North and Outer Schwyz. A great thank you to Jennifer Robinson for her important advice and support, as well as to Ute Lehrer and Richard Harris for their valuable comments.

REFERENCES

Architektengruppe Krokodil. 2012. *Glatt! Manifest für eine Stadt im Werden*. Zurich: Park Books.

Ascher, F. 1995. *Métapolis ou l'avenir des villes*. Paris: Jacob.

Aydalot, P., and A. Garnier. 1985. "Périurbanisation et suburbanisation: des concepts à definer." *DISP* 21 (80–1): 53–5. https://doi.org/10.1080/02513625.1985.10708447.

Bärtschi, H.-P. 1983. *Industrialisierung, Eisenbahnschlachten und Städtebau: die Entwicklung des Zürcher Industrie- und Arbeiterstadtteils Aussersihl*. Basel: Birkhäuser.

Bassand, M., and M. Schuler. 1985. *La Suisse, une métropole mondiale?* IREC, Rapport de recherche 54. Lausanne: EPFL.

Bassand, M., D. Joye, and M. Schuler, eds. 1988. *Les enjeux de l'urbanisation – Agglomerationsprobleme in der Schweiz*. Bern: ROREP, Lang.

Bätzing, W. 2015. *Die Alpen. Geschichte und Zukunft einer europäischen Kulturlandschaft*. 4. Munich: Beck.

Blanc, J.-D., and C. Luchsinger, eds. 1994. *Achtung: die 50er Jahre! Annäherungen an eine widersprüchliche Zeit*. Zurich: Chronos.

Brenner, N., ed. 2014. *Implosions/Explosions: Towards a Study of Planetary Urbanization*. Berlin: Jovis.

Brenner, N., and R. Keil, eds. 2006. *The Global Cities Reader*. New York: Routledge.

Brenner, N., and C. Schmid. 2014. "The 'Urban Age' in Question." *International Journal of Urban and Regional Research* 38 (3): 731–55. https://doi.org/10.1111/1468-2427.12115.

Brenner, N., and C. Schmid. 2015. "Towards a New Epistemology of the Urban?" *City* 19 (2–3): 151–82. https://doi.org/10.1080/13604813.2015.1014712.

Campi, M., F. Bucher, and M. Zardini. 2001. *Annähernd perfekte Peripherie. Glattalstadt and Greater Zurich Area*. Basel: Birkhäuser.

Coen, L., and C. Lambelet, eds. 2011. *L'Ouest pour horizon*. Schéma directeur de L'Ouest lausannois. Gollion: Infolio.

Corboz, A. 1988. *Stadt der Planer – Stadt der Architekten*. Zurich: VdF.

Corboz, A. 2000. "La Suisse comme hyperville." *Le Visiteur* 6: 112–29.

Crevoisier, O., J. Corpataux, and A. Thierstein. 2001. *Intégration monétaire et régions: des gagnants et des perdants*. Paris: L'Harmattan.

Cunha, A. and Schuler, M. 2001. "Métropolisation, changement de régime d'urbanisation et fragmentation de l'espace: enjeux de la gouvernance des agglomérations en Suisse," *Revue suisse de science politique* 7 (4): 119–26.

Diener, R., J. Herzog, M. Meili, P. de Meuron, and C. Schmid. 2006. *Switzerland: An Urban Portrait*. Basel: Birkhäuser.

Diener, R., L. Gunnarsson, M. Gunz, V. Jovanović, M. Meili, C. Müller Inderbitzin, and C. Schmid, eds. 2016. *Territory. On the Development of Landscape and City*. Zurich: Park Books.

Dümmler, P., C. Abegg, C. Kruse, and A. Thierstein. 2004. *Standorte der innovativen Schweiz. Räumliche Veränderungsprozesse von High-Tech und Finanzdienstleistungen*. Neuchâtel: Bundesamt für Statistik.

Eisinger, A. 2004. *Städte bauen: Städtebau und Stadtentwicklung in der Schweiz 1940–1970*. Zurich: gta Verlag.
Eisinger, A., and M. Schneider, eds. 2003. *Urbanscape Switzerland: Topology and Regional Development in Switzerland, Investigations and Case Studies*. Basel: Avenir Suisse, Birkhäuser.
ETH Studio Basel. 2010. *Metropolitanregion Zürich: Der Zürichsee als Projekt*. Zurich: Zürcher Handelskammer und Verlag Neue Zürcher Zeitung.
ETH Wohnforum 2010. S5-Stadt. *Agglomeration im Zentrum*. eBook. Accessed 24 June 2016. http://www.s5-stadt.ch/fileadmin/ebook/s5-stadt_ebook.pdf.
Fishman, R. 1991. "The Garden City Tradition in the Postsuburban Age." *Built Environment* 17: 232–41.
Forman, R. ed. 2008. *Urban Regions. Ecology and Planning Beyond the City*. Cambridge: Cambridge University Press.
Garnier, A. 1985. "Une région périurbaine suisse: Le Gros-de-Vaud." *DISP* 21 (80–1): 77–83. https://doi.org/10.1080/02513625.1985.10708452.
Garreau, J. 1991. *Edge City. Life on the New Frontier*. New York: Doubleday.
Glanzmann, L., S. Gabi, C. Kruse, A. Thierstein, and N. Grillon. 2006. "European Metropolitan Region Northern Switzerland: Driving Agents for Spatial Development and Government Responses." In *The Polycentric Metropolis: Learning from Mega-City Regions in Europe*, ed. P. Hall and K. Pain, 172–9. London: Earthscan.
Gunz, M., and C. Mueller Inderbitzin. 2008. *Thurgau: Projekte für die Stillen Zonen*. Zurich: Think Tank Thurgau, Niggli.
Hall, P., and K. Pain, eds. 2006. *Polycentric Metropolis: Learning from Mega-City Regions in Europe*. London: Earthscan.
Hamel, P., and R. Keil, eds. 2015. *Suburban Governance: A Global View*. Toronto: University of Toronto Press.
Harris, R. 2010. "Meaningful Types in a World of Suburbs." In *Suburbanization in Global Society*, ed. M. Clapson and R. Hutchison, 15–47. Bingley, UK: Emerald Group. https://doi.org/10.1108/S1047-0042(2010)0000010004.
Harrison, J., and M. Hoyler, eds. 2015. *Megaregion: Globalization's New Urban Form?* Cheltenham, UK: Edward Elgar. https://doi.org/10.4337/9781782547907.
Hermann, M. and Leuthold, H. 2003. *Atlas der politischen Landschaften*. Ein weltanschauliches Porträt der Schweiz. Zurich, vdf.
Hitz, H., C. Schmid, and R. Wolff. 1995. "Boom, Konflikt und Krise – ZürichsEntwicklung zur Weltmetropole." In *Capitales Fatales. Urbanisierung und Politik in den Finanzmetropolen Frankfurt und Zürich*, ed. H. Hitz, R.

Keil, U. Lehrer, K. Ronneberger, C. Author, and R. Wolff, 208–82. Zurich: Rotpunkt.

Hitz, H., C. Schmid, and R. Wolff. 1994. "Urbanization in Zurich. Headquarter Economy and City-Belt." *Environment and Planning. D, Society & Space* 12 (2): 167–85. https://doi.org/10.1068/d120167.

Jaggy, Y., and J.-B. Racine, eds. 2007. "Echelles et enjeux de la ville. Dossier." *Revue Économique et Sociale* 65 (4): 9–122.

Keil, R. 2013a. "Welcome to the Suburban Revolution." In *Suburban Constellations*, ed. R. Keil, 9–15. Berlin: Jovis.

Keil, R., ed. 2013b. *Suburban Constellations*. Berlin: Jovis.

Keil, R., and J.-P. Addie. 2015. "'It's not going to be suburban, it's going to be all urban': Assembling Post-Suburbia in the Toronto and Chicago Regions." *International Journal of Urban and Regional Research* 39 (5): 892–911. https://doi.org/10.1111/1468-2427.12303.

Krautzig, S., M. Gunz, and C. Mueller Inderbitzin. 2012. *Südliches Bodenseeufer: Projekt für eine urbanisierte Kulturlandschaft*. Zurich: Think Tank Thurgau, gta Verlag.

Kretz, S., and L. Küng, eds. 2016. *Urbane Qualitäten: ein Handbuch am Beispiel der Metropolitanregion Zürich*. Zurich: Edition Hochparterre.

Kretz, S., and K. Christiaanse. 2018, forthcoming. "Urbane Konstellationen: Städtebauliche Analysen in Zürich Nord und Ausserschwyz." In *Die Metropolitanregion Zürich: Urbanisierung, Städtebau und Urbanität*, ed. C. Schmid. Zurich: gta Verlag.

Lampugnani, V., M. Noell, G. Barman-Krämer, A. Brandl, and P. Unruh, eds. 2007. *Handbuch zum Stadtrand, Gestaltungsstrategien für den suburbanen Raum*. Basel: Birkhäuser.

Lefebvre, Henri. 2003 [1970]. *The Urban Revolution*. Minneapolis: University of Minnesota Press.

Lehrer, U. 1994. "Images of the Periphery: The Architecture of FlexSpace in Switzerland." *Environment and Planning. D, Society & Space* 12 (2): 187–205. https://doi.org/10.1068/d120187.

Lehrer, U. 2013. "FlexSpace – Suburban Forms." In *Suburban Constellations*, ed. R. Keil, 58–62. Berlin: Jovis.

Leresche, J.-P., D. Joye, and M. Bassand, eds. 1995. *Métropolisations: Interdépendances mondiales et implications lémaniques*. Geneva: Éditions Georg.

Marco, D., C. Schmid, C. Hirschi, D. Hiler, and J. Capol. 1997. *La ville: villes de crise ou crise des villes. Rapport scientifique final pour le Fonds national suisse de la recherche scientifique*. Geneva: Institut d'Architecture de l'Université de Genève.

McGee, T. 1991. "The Emergence of Desakota Regions in Asia: Expanding a Hypothesis." In *The Extended Metropolis: Settlement Transition in Asia*, ed. N. Ginsburg, B. Koppel, and T.G. McGee, 3–28. Honolulu: University of Hawaii Press.

Meili, A. 1941. *Landesplanung in der Schweiz*. Zurich: Neue Zürcher Zeitung.

Muri, G. 2018, forthcoming. "Topologien der Urbanität: Alltagswelten in Raum und Zeit." In *Die Metropolitanregion Zürich: Urbanisierung, Städtebau und Urbanität*, ed. C. Schmid. Zurich: gta Verlag.

Nüssli, R. 2015. "Auf dem Weg zu einer metropolitanen Regulation? Der Verein Metropolitanraum Zürich." *Geographica Helvetica* 70 (1): 11–25. https://doi.org/10.5194/gh-70-11-2015.

Nüssli, R. 2017. "Between Farming Villages and Hedge Fund Centres: The Politics of Urbanization in the Border Zone of the Metropolitan Region of Zurich." In *Old Europe, New Suburbanization? Governance, Land, and Infrastructure in European Suburbanization*, ed. N. Phelps, 207–36. Toronto: University of Toronto Press.

Nüssli, R., and C. Schmid. 2016. "Beyond the Urban-Suburban Divide: Urbanization and the Production of the Urban in Zurich North." *International Journal of Urban and Regional Research* 40 (3): 679–700. https://doi.org/10.1111/1468-2427.12390.

Oswald, F., P. Baccini, and M. Michaeli. 2003. *Netzstadt: Designing the Urban*. Basel: Birkhäuser.

Perlik, M. 2001. *Alpenstädte. Zwischen Metropolisation und neuer Eigenständigkeit*. Bern: Geographica Bernensia.

Perlik, M. 2010. "Leisure Landscapes and Urban Agglomerations – Disparities in the Alps." In *Challenges for Mountain Regions – Tackling Complexity*, ed. A. Borsdorf, G. Grabherr, K. Heinrich, B. Scott, and J. Stötter, 112–19. Vienna: Böhlau.

Phelps, N., A. Wood, and D. Valler. 2010. "A Postsuburban World? An Outline of a Research Agenda." *Environment & Planning A* 42 (2): 366–83. https://doi.org/10.1068/a427.

Phelps, N., and F. Wu, eds. 2011. *International Perspectives on Suburbanisation: A Post-Suburban World?* Basingstoke, UK: Palgrave Macmillan. https://doi.org/10.1057/9780230308626.

Racine, J.-B., and C. Raffestin, eds. 1990. *Nouvelle géographie de la Suisse et des Suisses*. Lausanne: Éditions Payot.

Salewski, C., and M. Michaeli. 2012. "Airport Corridor Zürich toevallig ontstaan." *Stedenbouw en Ruimtelijke Ordnung* 93 (6): 28–34.

Sassen, S. 1991. *The Global City: New York, London, Tokyo*. Princeton, NJ: Princeton University Press.

Schmid, C. 1989. *Der Urbanisierungsprozess in der Schweiz seit 1945*. Zurich: Diplomarbeit, Geographisches Institut der Universität Zürich.
Schmid, C. 1998. "The Dialectics of Urbanisation in Zurich: Global City Formation and Urban Social Movements." In *Possible Urban Worlds*, ed. INURA, 216–25. Basel: Birkhäuser.
Schmid, C. 2006a. "Global City Zurich: Paradigms of Urban Development." In *The Global Cities Reader*, ed. N. Brenner and R. Keil, 161–9. New York: Routledge.
Schmid, C. 2006b. "Theory." In *Switzerland: An Urban Portrait*, ed. R. Diener, J. Herzog, M. Meili, P. de Meuron, and C. Schmid, Vol. 1, 163–223. 3 vols. Basel: Birkhäuser.
Schmid, C. 2007. "Wie gross ist Zürich? Anmerkungen zu einer aktuellen Debatte." *Revue Économique et Sociale* 65: 67–82.
Schmid, C. 2011. "Henri Lefebvre, the Right to the City, and the New Metropolitan Mainstream." In *Cities for People, Not for Profit: Critical Urban Theory and the Right to the City*, ed. N. Brenner, P. Marcuse, and M. Mayer, 42–62. New York: Routledge.
Schmid, C. 2012. "Images of Urbanization in Switzerland." In *Auf Gemeindegebiet – On Common Ground*, ed. U. Görlich, and M. Wandeler, 168–72. Zurich: Scheidegger & Spiess.
Schmid, C. 2014. "Travelling Warrior and Complete Urbanization in Switzerland: Landscape as Lived Space." In *Implosions/Explosions: Towards a Study of Planetary Urbanization*, ed. N. Brenner, 90–102. Berlin: Jovis.
Schmid, C. 2015. "Specificity and Urbanization: A Theoretical Outlook." In *The Inevitable Specificity of Cities*, ed. R. Diener et al., 287–307. Zurich: Lars Müller.
Schmid, C. 2016. "The Urbanization of the Territory: On the Research Approach of ETH Studio Basel." In *Territory: On the Development of Landscape and City*, ed. R. Diener et al., 22–48. Zurich: Park Books.
Schmid, C. 2018, forthcoming. "Die Metropolitanregion Zürich: Urbane Konfigurationen und Paradigmen der Stadtentwicklung." In *Die Metropolitanregion Zürich: Urbanisierung, Städtebau und Urbanität*, ed. C. Schmid. Zurich: gta Verlag.
Schmid, C., and D. Weiss. 2004. "The New Metropolitan Mainstream." In *The Contested Metropolis: Six Cities at the Beginning of the 21st Century*, ed. INURA and R. Palosicia, 252–60. Basel: Birkhäuser.
Schubarth, C., and F. Auderset. 2004. *Glatt(t)alstadt: nouveau mot ou nouvelle ville?* Fribourg: Université de Fribourg.
Schuler, M., M. Perlik, and N. Pasche. 2004. *Nicht-städtisch, rural oder peripher – wo steht der ländliche Raum heute? Analyse der Siedlungs- und*

Wirtschaftsentwicklung in der Schweiz. Bern: ARE, Bundesamt für Raumentwicklung.

Schuler, M., P. Dessemontet, C. Jemelin, A. Jarne, N. Pasche, and W. Haug. 2006. *Atlas des mutations spatiales de la Suisse / Atlas des raumlichen Wandels der Schveiz.* Neuchatel: Office fédéral de la Statistique / Zurich: Verlag Neue Zürcher Zeitung

Schwick, C., J. Jeager, R. Bertiller, and F. Kienast. 2012. *L'étalement urbain en Suisse – impossible à freiner? Urban Sprawl in Switzerland – Unstoppable?* Bern: Haupt.

Scott, A. 1998. *Regions and the World Economy*. Oxford: Oxford University Press.

Seifert, J., F. Murat, F. Bühler, and R. Blödt, eds. 2006. *Beyond Metropolis: Eine Auseinandersetzung mit der verstädterten Landschaft*. Sulgen: Niggli.

Sieverts, T. 2000 [1997]. *Cities without Cities: An Interpretation of the Zwischenstadt*. London: Spon Press.

Soja, E.W. 1992. "Inside Exopolis: Scenes from Orange County." In *Variations on a Theme Park*, ed. M. Sorkin, 94–122. New York: Noonday Press.

Soja, E.W. 2000. *Postmetropolis: Critical Studies of Cities and Regions*. Oxford: Blackwell.

Soja, E.W. 2010. "Regional Urbanization and the End of the Metropolis Era." In *The New Blackwell Companion to the City*, ed. G. Bridge and S. Watson, 679–89. Cambridge, MA: Blackwell.

Soja, E., and M. Kanai. 2006. "The Urbanization of the World." In *The Endless City*, ed. R. Burdett and D. Sudjic, 54–69. London: Phaidon.

Storper, M., and A. Scott. 2016. "Current Debates in Urban Theory: A Critical Assessment." *Urban Studies* 53 (6): 1114–36. https://doi.org/10.1177/0042098016634002.

Studer, S. 1985. *Nationalstrasse – Nationalstrafe, oder: Die Demokratie bleibt auf der Strecke. Macht und Ohnmacht im schweizerischen Nationalstrassenbau*. Zurich: Rotpunktverlag.

Taylor, L., and P. Hurley, eds. 2016. *A Comparative Political Ecology of Exurbia: Planning, Environmental Management and Landscape Change*. Springer. https://doi.org/10.1007/978-3-319-29462-9.

Thierstein, A., C. Kruse, L. Glanzmann, S. Gabi, and N. Grillon. 2006. *Raumentwicklung im Verborgenen. Die Entwicklung der Metropolregion Nordschweiz*. Zurich: Verlag Neue Zürcher Zeitung.

Thierstein, A., T. Held, and S. Gabi. 2005. "Zurich/Glattal: City of Regions." In *Urbanscape Switzerland. Typology and Regional Development in Switzerland*, ed. A. Eisinger and M. Schneider, 297–331. Basel: Birkhäuser.

Veltz, P. 1996. *Mondialisation, villes et territoires. L'économie d'archipel*. Paris: PUF.

Walter, F. 1994. *La Suisse urbaine 1750–1950*. Geneva: Éditions Zoé.
Wong, T., N. Hanakata, A. Kockelkorn, C. Schmid, and R. Sullivan. 2017. "Multilayered Patchwork Urbanization in Tokyo, Hong Kong, Paris and Los Angeles." Working paper, ETH Zurich.
Young, D., P. Wood, and R. Keil, eds. 2011. *In-between Infrastructure: Urban Connectivity in an Age of Vulnerability*. Kelowna, BC: Praxis.

chapter five

The Morphology of Dispersed Suburbanism: The Land-use Patterns of the Dominant North American Urban Form

PIERRE FILION

The post-World War II period has witnessed the appearance and rapid dissemination across North America of a novel urban form – the dispersed suburb – marking a radical break from precedent urban patterns. Dispersed suburbanism is characterized by heavy reliance on the automobile, low overall density, land-use specialization, and a scattering, in opposition to clustering, of employment, retailing, services, and institutions (hence the "dispersed" label) (Filion, Bunting, and Warriner 1999). The spread of dispersed suburbanism has been so extensive that a majority of North Americans now reside and work within such environments (Cervero 2013, 22–6).

The chapter seeks to advance the understanding of dispersed suburbanism by investigating its morphology. The focus is on prominent features defining this urban form: the role expressways and arterials play in structuring this type of environment and the nature and distribution of specialized land uses. The method relies on observation, measurements, and analyses of morphological features to investigate how suburban land use is organized and how this organization is achieved. The interpretation of the findings will highlight factors shaping the morphology of dispersed suburbanism and assuring its ongoing reproduction: the ubiquity of the car, a levelling of accessibility potentials, and a specialization of land use.

After attempting to derive as much understanding of the dispersed suburb as possible from morphological features, the chapter ties these findings to upstream processes – planning and the role of developers. It also examines circumstances explaining the flourishing in the present neoliberal societal context of an urban formula devised in the Fordist/Keynesian demand-side economic policy climate

The area under study is a slice of the Toronto region suburban belt: the continuously built-up area occupying the southern portion of York Region. While largely characteristic of the North American dispersed suburb, this sector also presents some specificities, such as density and public transit use in excess of the continental suburban norm along with less administrative and hence planning fragmentation. Still, it remains well suited to the exposure of the main morphological features of dispersed suburbanism.

Analysing Suburban Morphology

Their morphology reveals much about cities: the importance given to different activities, their evolution, and their dynamics (especially interrelations between land use and transportation). All aspects of cities are tied to their use of space and interconnections between different functions, which makes it possible to deduce from observation of their built form how cities function and are reproduced. Of course, urban morphology can be an object of interest in its own right, as it lends itself to research exploring its taxonomy and evolution and the comparison of urban forms (Scheer 2010). Urban morphology has evolved into a field of inquiry proposing methodologies that bring rigour to the observation of urban form, thereby making it possible to disclose regularities and laws (Conzen 2004; Levy 1999; Moudon 1997). These can concern urban morphology in general or be specific to one category of cities, or one growth period (Vance 1990). The chapter focuses on the type of urban development that is most common across North America and gaining ground elsewhere: the dispersed suburb. The study presented here uses urban morphology methods as instruments to advance understanding of dispersed suburban development (Whitehand and Carr 2001).

Over the last decades much reflection has been given to the post-World War II suburban morphology, which rapidly came to dominate urban development in North America while gaining importance in other parts of the world. Discussions of this urban form came from different perspectives. Some writing was all-out critical of post-war suburbanism. Perhaps most antagonistic was James Kunstler (1993), who in the *Geography of Nowhere* lambasted the post-war North American suburb for its lack of place-making capacity responsible for a standardized environment bereft of identity. At the other extremity of the spectrum were apologists for the suburb, celebrating its characteristic aesthetic. Robert Venturi was among the first to pay tribute to the architecture and landscapes of the post-war suburb (Venturi, Scott Brown, and Izenour 1977). Rather than either criticizing or celebrating it, others have attempted to further understand this urban phenomenon by exploring its geography, spatial organization, or rapport with the evolving political economy.

Inspired by the urban structure of Los Angeles, Dear and Flusty (1998) have put forth an urban model that highlights recent forms' ability to accommodate side-by-side land uses that are profoundly different, when not outright incompatible. To underscore the randomness of its land-use distribution, these researchers label their urban model "keno capitalism" after the keno gambling game. Other researchers have concentrated on the factors underpinning post-war suburban development and on the nature of this form of urbanization: What land uses decentralize? Where do they locate? What form do they take in the dispersed suburb? For example, the FlexSpace approach documents the location and architecture of Swiss suburban developments (Lehrer 1994). Recently, much attention has gravitated towards the *Zwischenstadt*, translated as "in-between city," perspective (Keil and Young 2011; Sieverts 2003). Researchers subscribing to this approach depict the in-between city as wedged between traditional forms of urbanization and rural areas. They further document the extreme diversity of land uses found within these urban forms as well as their heavy reliance on the automobile. While researchers acknowledge the North American origin of many features of the in-between city as well as the presence in this continent of the fullest expression of this urban form, they also describe its variants in different parts of the world (Schmid 2012).

The dispersed suburb perspective attempts to advance these reflections on the post-war suburb. It takes the view that the urban form of the dispersed suburb along with its interrelation with transportation patterns are a major source of influence on the expectations and behaviour of the public and thereby contribute to the reproduction of this urban pattern. Just as Derrida (1974) claimed that "there is nothing outside the text" and that it is possible to interpret a text without reference to factors that are external to it, one can assume somewhat more guardedly that it is possible to gain much understanding of an urban pattern by investigating its morphology and journeys. While the chapter adheres to this methodological orientation, it also heeds broader societal influences on dispersed suburbanism.

To grasp the defining features of the accessibility pattern of dispersed suburbanism, it is best to compare it with accessibility distributions within the highly centralized form that preceded it as the dominant urban model. The pre-war urban form was configured by public transit, downtown-focused, hub-and-spoke networks. Resulting accessibility gradients peaked at the downtown and declined sharply as distance from this sector grew. Such accessibility patterns favoured intense concentrations of activities in downtowns, which triggered the invention of the skyscraper to optimize land use therein.

Generalized reliance on the car has yielded an accessibility pattern and urban form at the antipodes of the public transit induced, centralized, urban structure. First, because of the traffic congestion it causes and the space it consumes, the automobile is inimical to large concentrations of activities as found in downtowns. The amount of space it requires counters the organizing principles at the heart of the existence and growth of downtowns. A key ingredient of successful downtowns is the existence of powerful, pedestrian-based interactions between mutually dependent activities such as retailing and offices, and hotels and convention centres. The car occupies downtown space at the expense of the range of activities in this district and pedestrian-based interactions among these activities (Filion 2007). Automobiles are agents of decentralization and scattering of activities. Automobile-dominated urban environments do not produce accessibility peaks of an amplitude comparable to that found in urban areas shaped by public transit. Accessibility gradients associated with the car are flat by comparison. Instead of one unchallenged focal point, they give rise to multiple but comparatively modest protuberances, which are found at expressway interchanges and exits (Bottles 1987; Muller 2004).

The generalization of car use is also responsible for an overall rise in accessibility levels, which makes it possible to consume far more urban space per person or per economic activity. In addition, with the levelling of accessibility gradients, the location of a household or activity loses importance from an accessibility point of view. Other factors of localization then fill the void and assume more prominence. Such is the case of a preference for proximity to land uses that are similar to one's own. Combined with land-use possibilities unleashed by the capacity automobiles have of accessing scattered origins and destinations, the preference for land-use homogeneity results in large mono-functional zones (Rowe 1991). Ironically, another land-use impact of generalized automobile use is a consequence of intolerance on the part of residents towards high levels of vehicle circulation. One characteristic of dispersed suburbanism is the sheltering of residential areas from traffic noise and hazards (Appleyard 1980; Charmes 2010; Wang and Smith 1997).

The above paragraphs have focused on the impact of generalized automobile use on dispersed suburbanism. They could just as well have described how primarily low-density and mono-functional built forms promote reliance on the automobile, for the relation between transportation and land use in dispersed suburbanism (as in other urban forms) is bidirectional. To be sure, other dynamics fuel dispersed suburbanism, as in the case of consumer values and preferences that both are shaped by the prevailing urban environment and influence the production of this environment. But all these other dynamics are to some extent connected to the transportation and land-use interrelation in dispersed suburbs.

Findings from the morphological analysis presented in the chapter will reveal how the spatial organization of suburban dispersion echoes automobile-induced accessibility gradients and reactions to the negative externalities of this mode of transportation. More generally, the analysis will address the following questions: What morphological configurations does the dispersed suburb take? What are the components of its morphology? How do they relate to each other? What are the regularities (or laws) that are revealed by the analysis?

The Object of Study: The Southern Portion of York Region

The suburban sector under study covers the area that runs from the northern border of the city of Toronto to the end of the continuously built-up area as of 2009 (see Figure 5.1). It includes three local municipalities – Vaughan, Richmond Hill, and Markham – which constitute the southern portion of York Region, a second-tier regional municipality. It is important to note planning philosophy differences between the local municipalities. While Vaughan and Richmond Hill have subscribed over past decades to conventional patterns of suburban development, since the mid-1990s Markham has actively promoted New Urbanism-style development (Langlois 2010). The 268 km^2 band of suburban development under investigation registers a 2011 population

Figure 5.1 Study area: built-up super blocks in the southern portion of York Region, Canada. Drawn by the author on Google Earth® aerial imagery

of 631,791, which constitutes 11 per cent of the Toronto census metropolitan area population.

The study area was developed from the late 1970s until now, illustrating patterns of dispersed suburbanism in place over the last forty years. Of late, the population growth rate has moderated, as it now approximates that of the census metropolitan area. Individual incomes are slightly less than those of the census metropolitan area, but household incomes are substantially higher, a function of much larger households in the study area.

A comparison of modal shares in the southern portion of York Region with that of the City of Toronto and the Greater Toronto and Hamilton Area, an extended Toronto-focused region with a population of 6,574,140 (approximately 1 million in excess of that of the Toronto census metropolitan area), reveals the automobile orientation of the study area. It scores highest in terms of driver and passenger modal shares and records the lowest local transit and walking and cycling statistics. Local transit modal shares are more than three times less than those of the City of Toronto, which itself is partly composed of sectors with a dispersed suburbanism configuration but also contains a public transit- and pedestrian-conducive downtown and inner city (TTS 2006).

Given its period of development and its automobile reliance, there are good reasons to expect conformity between York Region and North American dispersed suburban development norms. However, one cannot ignore the possibility of some variations from the North American model stemming from circumstances specific to the study area. Residential density is higher in Toronto outer suburbs (those developed after 1971) than North American averages for this generation of suburbs (Filion et al. 2004). In a similar vein, although low by Toronto region standards, public transit modal shares in the study area remain well above North American suburban levels. Finally, the presence of only three local municipalities, themselves overarched by a regional administration, affords more planning coordination capacity than is found in more fragmented suburban jurisdictions, a common occurrence across the continent. Among other things, we can expect a better integration of road networks than in the suburban realms of many other North American metropolitan regions.

Methodology

The presentation of the methodology defines the scale of resolution of the morphological analysis. Urban morphology includes a remarkably wide spectrum of urban features, from architectural details and street furniture to the shape of the overall urban envelope of a metropolitan region. Aspects of the urban morphology methodological tool kit (Moudon 1997) relied upon in the chapter are determined by the objectives of the study: the identification of the main defining land-use features of dispersed suburbs and the way they interact. The chapter's concern for the structure of the dispersed suburb and the presence and distribution therein of different land uses calls for a coarse-grain approach centring on major road networks and land-use zones. It concentrates on main land-use categories: residential, employment, and retailing. The chapter does not heed subcategories within these land uses, with the exception of high- and low-density housing when separators between potentially conflicting land uses are considered. The identification and measurement of these land-use features rely on 2009 aerial photographs using Google Earth Pro images and software. The basic units of morphological description (Batty 1999, 2) are the super-grid/super-block structure and functionally specialized zones.

There are three steps in the methodology. The first concentrates on expressway and arterial road patterns and their structuring effect on land use. It describes the super-grid/super-block layout defined by expressways and arterials, along with the effects of this key feature of dispersed suburbanism on accessibility and land use. The next step is concerned with the land-use components of the study area. It identifies types of land use along with their respective importance. A dominant theme of this second step is the relation between zones of specialized land use and the super-grid/super-block structure. The final step examines techniques used to achieve land-use specialization in the context of the dispersed suburb. It itemizes the different methods relied upon to separate land uses and computes differences in the extent to which each method is used.

Structure and Components of the Morphology of Dispersed Suburbanism

Figure 5.2 illustrates the super-grid/super-block configuration of the sector under investigation. Its area is segmented into seventy-nine super blocks defined by the expressway and arterial network. There are irregularities in the super-grid pattern, the outcome of historical factors, physical geography, and the presence of expressways. These irregularities result in wide variations in the shape and size of super blocks. Their mean size is 3.25 km^2, ranging from 0.56 to 8.85 km^2. A 1.84 standard deviation coefficient reflects the wide distribution of super-block sizes.

Virtually all super blocks exhibit modified grid and curvilinear road patterns with limited connections to the super grid. Table 5.1 presents the number of such connections per super block. To place its statistics into perspective, the area-number of connections to the super-grid ratio of study area super blocks is compared with that of a City of Toronto inner-city grid-iron district approximating the area of the average super block (the surface of the inner-city district is 3 km^2). The average-sized super block in our study is connected to the super grid by 10.1 roads, which amounts to one connection per 0.322 km^2. In contrast, fifty-three streets link the inner-city district to its surroundings, resulting in one connection per 0.057 km^2. The low standard deviation coefficient reported in Table 5.1 for the area super-grid connections ratio of study area super blocks indicates a great deal of similarity in the relation between their local street network and the super grid. There is one outlier among the seventy-nine super blocks: a mostly grid-patterned New Urbanism community. Revealingly, its area connections

Figure 5.2 Southern portion of York Region, Canada: super blocks. Drawn by the author on Google Earth® aerial imagery

to the super-grid ratio approach that of the inner-city district used for comparative purposes.

The adoption of such a street pattern in study area super blocks could be interpreted as the result of efforts to safeguard residential areas from high traffic volumes generated by dispersed suburban configurations. To verify the validity of this possibility, Table 5.1 separately presents findings from residential and employment super blocks (super blocks are considered to be mono-functional when their second use occupies less than 10 per cent of their area). One would expect far fewer connections to the super grid in residential than in employment super blocks. However, there is surprisingly little variance in the area-connections to the super grid ratios of the two categories of super blocks. So the rationale for the observed street configuration must rest elsewhere. The most likely explanation is the adoption across all dispersed suburban land uses of a similar road pattern based on a traffic speed and volume hierarchy: expressways, arterials, collectors, modified grids, crescents, and cul-de-sacs (Groth 1981; Southworth and Ben-Joseph 1995). Such an organization implies reliance on collectors to link lower-order to higher-order roads – hence limited connections of super-block streets to the super grid. Both residential and employment zones feature modified grids, crescents, and even cul-de-sacs, but as expected at different scales to accommodate different building sizes (e.g., industrial plants versus single-family homes).

Table 5.1. Southern portion of York Region, Canada, super blocks: number of roads connecting them to the super grid

	Area (km²)	Number of roads connecting to the super grid	Area (km²)/ number of roads	Standard deviation of area (km²)/number of roads
Super blocks (79) Total	256.38	797	0.322	
Super blocks (79) Mean	3.25	10.1		0.138
Employment land use Super blocks (15) Total*	34.18	95	0.360	
Employment land use Super blocks (15) mean*	2.28	6.33		0.183
Residential land use Super blocks (42) total*	162.17	484	0.335	
Residential land use Super blocks (42) mean*	3.86	11.52		0.124

* Mixed-use super blocks (those where a second land use occupies 10 per cent or more of the area) are not included as a distinct category.

Source: Author's calculations

When we examine land use, the dominant trend to emerge from the study is a tendency for super blocks to be entirely occupied by a single land use. Table 5.2 indicates that 78.5 per cent of super blocks can be classified as mono-functional. Residential super blocks represent 54.4 per cent of all super blocks and employment super blocks account for 22.8 per cent of the total. Among employment super blocks most of the surface is taken by small one-storey facilities accommodating industries, warehouses, and offices.

There is no large factory, such as an automobile assembly plant, in the study area. Only one super block is entirely occupied by retailing. The highest proportion of the 21.5 per cent of functionally mixed super blocks combine residential and retail zones. The next highest proportion of multi-zone super blocks hosts residential and employment zones. There is only one super block occupied by both employment and retail zones. Finally, two super blocks in the study area present unique land uses: a theme park along with a residential zone, in one case, and a small-plane airport with an employment zone, in the other.

Table 5.2. Southern portion of York Region, Canada, super blocks: land use

Super-block land use	Number	Per cent
Residential	43	54.4
Employment	18	22.8
Retail	1	1.3
Total mono-functional*	62	78.5
Residential-employment	6	7.6
Residential-retail	8	10.1
Employment-retail	1	1.3
Other: Theme park-residential, Airport-employment	2	2.5
Total more than one land use**	17	21.5
Total all super block	79	100

* Super blocks where the second land use occupies less than 10 per cent of their area
** Super blocks where the second land use occupies 10 per cent or more of their area
Source: Author's calculations

The high degree of coincidence between mono-functional zones and super blocks could lead to the conclusion that expressways and arterials represent important separators of land uses. Such a conclusion is valid to some extent, as expressways and arterials do separate super blocks occupied by different land uses in seventeen instances. But two circumstances contribute to curb the extent of the land-use separator role of expressways and arterials in the study area. There are instances where, despite the presence of different land uses in two super blocks, both sides of an arterial that separates these super blocks are occupied by a similar land use (a minority land use in at least one of the two super blocks). However, such a situation has been observed in only three cases. With a much greater limiting effect on the degree to which expressways and arterials divide land uses is the tendency for super blocks with similar land uses to be adjacent to each other. Hence the existence of super zones, which combine several contiguous super blocks with similar functions. Much of the surface of the study area is indeed covered by residential and employment super zones (see Figure 5.3). Table 5.3 indicates that a majority of super blocks abut other zones with a similar function, and that only three of them are entirely surrounded by super blocks with functions that are different from their own.

As land-use specialization represents a predominant feature of dispersed suburbanism, methods of separating land uses play a key

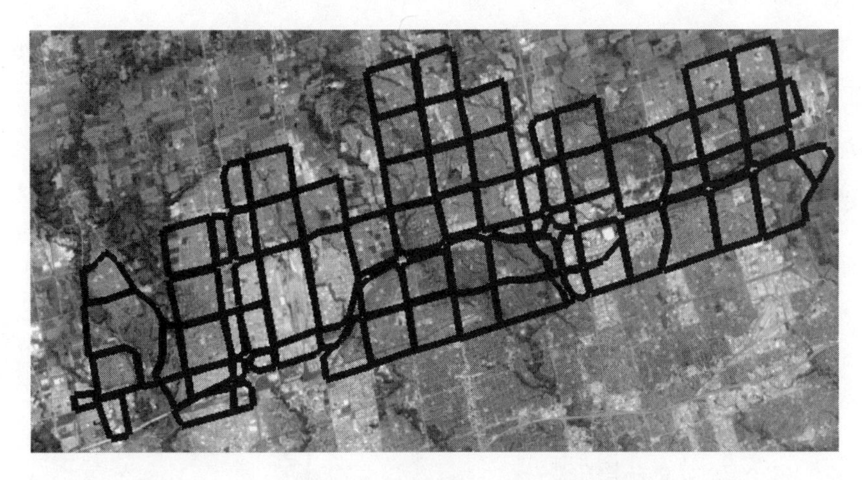

Figure 5.3 Southern portion of York Region, Canada: dominant land use in super blocks. Light grey shading refers to residential areas, medium grey shading to employment areas, and dark grey shading to other areas. Drawn by the author on Google Earth® aerial imagery

Table 5.3. Dominant land uses of super blocks surrounding a given super block (excluding undeveloped super blocks)

Super blocks surrounded by other super blocks with dominant land uses that are	Number	Per cent
All the same as their own dominant land use	42	53.2
The same as their own dominant land use except one	22	27.8
The same as their own dominant land use except two	9	11.4
The same as their own dominant land use except three	3	3.8
All different	3	3.8

Source: Author's calculations

role in the production of this urban pattern (Kunstler 1993; Marshall 2000). To understand the morphology of the dispersed suburb, it is thus important to consider the different forms land-use separators take. We have seen the part played by expressways and arterials, but what of methods used to divide zones within super blocks? The study has identified 298 individual land-use separators between different land uses inside super blocks (see Table 5.4). Note that these separators

Table 5.4. Land-use separators within super blocks*

Land uses	Housing backing on different land Uses	Street	Green space	Railway	Green cordon	Housing backing on different land uses and green space**	Street and green space**	Railway and green space**	Total
Retail-residential	109 (56.8%)	36 (18.8%)	24 (12.5%)	0	5 (2.6%)	4 (2.1%)	7 (3.7%)	7 (3.7%)	192 (100%)
Industrial-residential	15 (26.8%)	8 (14.3%)	17 (30.4%)	4 (7.1%)	0	1 (1.8%)	4 (7.1%)	7 (12.5%)	56 (100%)
High-density residential; low-density residential	15 (30%)	9 (18%)	11 (22%)	0	2 (4%)	3 (6%)	6 (12%)	4 (8%)	50 (100%)

* Includes all the different separators between two land uses. When a given separator is divided by a street or any other obstacle, it is counted as two separators.

** Green space here can refer to a large expanse or a narrow cordon.

Source: Author's calculations

concern both zones occupying 10 per cent or more of the surface of a super block and smaller zones. Only separators between retail and residential, employment and residential, and high-density and low-density residential land uses are considered. Accordingly, juxtaposed land uses with minimal potential for tensions, such as school and residential and employment and retail, are left out. All identified separators inside super blocks fall into eight categories: back to back, street, green space, railway, green cordon (i.e., narrow band), back to back and green space, street and green space, and railway and green space.

Not surprisingly, the retail-residential land use juxtaposition is the one that is associated with the largest number of separators. Retailing comes in different formats, occupying areas with vastly different sizes. Retail zones can be so tiny that they accommodate a single small store or so vast that they encompass regional shopping malls or massive power malls with their respective seas of parking. Most retail-residential separators in the study involve small retail zones. The most common form of separation between retail and residential land use is back to back, where houses back onto retail areas (generally parking lots or the back of stores). The second mode of retail-residential land-use division is the street, where retail occupies one side and housing the other. All other forms of division between retail and residential land uses involve green space: green space cordons or green space of various expanses associated with other separators. Together, however, these other forms amount to less than 10 per cent of retail-residential land-use dividers. There is less back-to-back separation in the case of juxtaposed employment and residential zones and more reliance on green space on its own or associated with other forms of separation. This is to be expected, given the high potential incompatibility between these two land uses. Railways also figure as a separator between employment and residential zones, which is also predictable, given the reliance of some manufacturing and warehousing on railways. Back-to-back and street separation account for nearly half the dividers between high- and low-density residential land uses, the last category of juxtaposed zones found in Table 5.4. All other forms of separation between these two land uses involve green space on its own or in conjunction with other separators.

Lessons from the Observation and Analysis of Dispersed Suburbanism

The present section examines the tight association between the predominance of automobile transportation, the road and highway system evolved to accommodate this mode, and land-use distributions in the dispersed suburb. We have seen that the super grid and automobile dependence together structure the land use of dispersed suburbanism. Just like the grid provided the organizing principle of inner-city land use, the super grid is the chief structuring agent of the dispersed suburb (Southworth and Ben-Joseph 1995, 2003; Southworth and Owens 1993 on how the grid disciplines the Manhattan urban form, Koolhaas 1978).

It is important to distinguish macro from micro accessibility in the interpretation of observed morphological patterns. Macro accessibility pertains to the relatively levelled accessibility gradients of dispersed suburbanism, whereas micro accessibility refers to variations at a local level, such as the difference between a location abutting an arterial and another located deep within a super block. Macro accessibility thus operates at the scale of the dispersed suburb in its entirety, while micro accessibility influences locations within super blocks. The macro accessibility of dispersed suburbanism introduces considerable randomness in the location of different types of activities. From an accessibility perspective any land use could have been located in any part of the study area. Indeed, if some employment areas are close to expressways, so are some residential areas. Likewise, major retail concentrations tend to be near expressway exits, but there are exceptions to the rule. True, it is possible to detect the influence of the presence of railways on the location of some employment areas. Marshalling yards are at the centre of the main concentration of industrial and warehousing facilities, and other zones containing such activities are situated close to railway lines.

The effects of variations in accessibility are more evident at the micro scale, especially in the case of retail. Virtually all retail fronts on arterials, which is not surprising, given that in such cases visibility is closely associated with accessibility. The impact of this pattern is especially clear in predominantly residential super blocks. In these areas nearly all retail is found at the edge of super blocks, whose interiors are occupied by housing and ancillary uses such as schools. The inside of residential super blocks, with their curvilinear street configuration, indeed discourages through traffic and provides oases from the noise and hazards of heavy traffic while being within easy automobile reach of the super grid.

As macro accessibility does not feature prominently in location decisions, other factors of location take over. Findings point to the importance of land-use compatibility, which is responsible for specialized land use and its large-scale projection into super zones. All else being equal, land uses locate close to other land uses with similar functions. It is the accretion of such decisions that leads to the emergence of super zones. The land-use aggregation pattern is further fuelled by heavy reliance on the car, making it difficult for functions such as large-scale retail or office space, which are not otherwise sources of negative externalities, to coexist with housing. It is the high traffic levels they generate that make these land uses incompatible with residential areas.

Levelled accessibility gradients and the ubiquity of the car are mirrored in the absence of large, multi-use concentrations of activities in the study area (Lang 2003). Such a situation derives from difficulties in achieving pedestrian-based synergies. We have seen that the existence of such synergies is essential to the tight clustering of interdependent activities. The absence of large, multifunctional centres is also a function of land values shaped by accessibility distributions within dispersed suburban environments. In relatively flat accessibility settings there is little incentive to pay higher land costs associated with large concentrations of activities rather than locate in one of the multiple other locations offering equivalent accessibility potentials. The development trajectory of the dispersed suburb is therefore not one of further intensification of existing zones, but one of replication. Instead of adding floors to an existing shopping mall, a new one is created in a sector under development. The same is true of other land uses. This type of development is compatible with the car-dependent, super-grid transportation system. The operation of such a system best approximates optimality when origins and destinations are scattered. A scattered pattern assures equivalent traffic flows in different directions and on different parts of the network. One downside of this type of development is the highly standardized nature of its landscape. The resulting lack of diversity and complexity is mirrored by the reliance on only three scales of observation (nested hierarchies in the urban morphology lexicon) in the study: specialized zones, super blocks, and super zones (Conzen 2004, 60–77). It also creates an environment that is inimical to non-automobile forms of transportation. Distances are generally too long for walking, as is often the case for cycling. What is more, heavy traffic on arterials imperils cyclists. With regard to public transit, this mode performs poorly when origins and destinations are dispersed (Cervero and Kockelman 1997). Efficient services require a concentration of destinations and origins posting sufficient density.

Efforts to depart from the conventional dispersed suburbanism model have not attained their land use and transportation objectives

(Langlois 2010). Cornell, a New Urbanism community east of Markham, has attracted few stores to its street-oriented commercial centre (Grant and Perrott 2011). Despite New Urbanism aspirations to provide pedestrian-friendly environments, opportunities for functional walking within Cornell are therefore limited. Moreover, at 91.6 per cent the Cornell automobile, van and truck driver, and passenger modal shares are substantially higher than in the study area average of 85 per cent and, at 2.6 per cent, reliance on public transit is more than 6 per cent lower (TTS 2006). Cornell illustrates how the dynamics of the super grid can override attempts at creating distinct journey patterns within super blocks.

Although not the object of a systematic comparison here, it appears that, as expected, some specificities singularize the study area in the universe of North American dispersed suburbanism. There is more development continuity in Toronto suburbs than in many other North American locations, owing to the effect of planning and the tendency for suburban growth to follow infrastructure extensions (road, water, and sewer) north of the existing urbanized perimeter (Filion et al. 2010). The density of the continuously built-up outer suburb (erected since 1971) of Toronto is highest among the similarly defined outer suburbs of a sample of fifteen North American metropolitan regions with a broadly comparable population (Filion et al. 2004). Finally, the presence of a regional government has provided more consistency in the layout of the super grid than in more administratively fragmented North American dispersed suburbs. Still, there is no reason to believe that these particularities set the study area apart as regards the dynamics of dispersion. Densities remain too low to support quality public transit services and the study area shares the land-use specialization, scattering of activities, and road hierarchy defining North American dispersed suburbanism.

Planning, Development, Neoliberalism, and Dispersed Suburbanism

In order to relate the findings to upstream processes and societal contexts, it is now time to lift the analytical bell jar that up to this point has isolated the morphological study from external factors of influence. More specifically, this section connects observations on the morphology of the study area to the planning and development processes responsible for the existence of this urban form and explores its relations to the prevailing neoliberal societal environment.

Regularities revealed by the morphological analysis attest to the existence of conventions guiding the development process, and thus to the role of planning. It is therefore important to examine the form of planning that has been at work in the study area. Planning did not operate as it does in those master-planned communities that are structured around different orders of strategically located multifunctional centres. Planning assures that different land uses in such communities relate to local and community centres and to the dominant regional centre. In this sense, the entire community is organized around a single, integrative concept: the hierarchy of centres. In the Toronto region such a planning formula is found in the Don Mills, Erin Mills, Meadowvale, and Bramalea master-planned districts (Sewell 1993).

Over its period of development there was no such overarching concept guiding the planning of the study area. Planning consisted, rather, in the delineation and allocation, as the super grid was extended, of zones according to foreseen demand for different functions. The upshot is the distribution of zones with specialized land uses observed in the morphological analysis. Unlike the situation prevailing in master-planned communities, zones are not located in relation to each other according to a predetermined planning vision. In commonplace dispersed suburbanism, as encountered in the study area, interrelations between zones are a function not of juxtaposition or proximity to specific destinations such as different orders of centres, but rather of their connection to the super grid. The hierarchical organization of space found in master-planned communities makes way for mosaic land-use compositions in conventional dispersed suburbanism. The management of the development of the study area thus was guided by a coarse form of planning, concerned primarily with the allocation of specialized zones. An associated planning role was that of preventing tensions between incompatible land uses. We have seen the part played by the super grid in this regard. We have also noted the dependence on green spaces among other means to separate land uses. By relying on the super-grid structure and the abundant availability of space in the dispersed suburb to keep different land uses apart, planning minimizes the need for efforts at integrating and blending land uses.

Once zoning is in place, developers submit plans for an entire zone or portion thereof. Depending on the size of zones, such plans can cover a super block in its entirety. As developers specialize in a single type of development, their activities are typically confined to one category of zones. In Ontario developers are responsible for the building of infrastructures within their subdivisions and, through development charges, provide funding for arterials and other municipal infrastructures. These are the planning and development mechanisms that undergirded the production of the built environment of the study area.

Turning now to the relation between the dispersed suburb morphology and the broad societal context, one cannot escape addressing the question of how a suburban model bequeathed by Fordism/Keynesianism, and that conformed to the demand-side economic policies of the time, persists and prospers in the present neoliberal era. From the late 1940s until the mid-1970s dispersed suburbanism was pieced together as a catalyst for the consumption of durable goods at a time of accelerated middle-class expansion (Marglin and Schor 1990; Moulaert, Swyngedouw, and Wilson 1988). Ongoing reliance on this model in a prolonged period of middle-class shrinkage and public-sector financial stringency requires explanations. New suburban development is funded by high- and middle-income households. More precisely, most of the money for new housing purchases is a function of the capacity of households to take on mortgages. Ultimately, much of the dispersed suburbanism development process therefore rests on household debt, which allows it to keep on expanding, at least for a time, in the face of the financial woes confronting the middle class in a neoliberal age (Walks 2013). This debt contributes to funding houses, infrastructures within residential super blocks, and the super grid via development charges rolled in the price of the home. Housing purchasing power in the study area is related not so much to individual income (below the metropolitan mean) as to the large size of households, which allows the pooling of incomes. Parallel to housing-based financial mechanisms are those associated with commercial development (retail and employment). New dispersed suburbs are well positioned to attract retail catering to their residents as well as workplaces seeking the relatively low taxes and development costs afforded by plentiful accessible space and heavy local and regional government reliance during the growth phase of these suburbs on revenues originating from development charges. Compatibility with neoliberalism relates to the mostly private financing of the development of the dispersed suburb along with the restrained planning interventions guiding its development (Blais 2011; Hackworth 2006; Harvey 2005).

The effect on dispersed suburbanism of the slow-growth economy over the neoliberal phase and the attendant middle-class shrinkage is felt in older, filtering-down, dispersed suburban areas. These areas must confront simultaneously the loss of their industrial base, a casualty of de-industrialization, and falling incomes as residents retire and are eventually replaced by poorer households (Hulchanski 2010; Lucy and Phillips 2000). Such a progression happens in a context where development charges yields are reduced to a trickle, while maintenance costs of aging infrastructures escalate. Getting old is not kind to dispersed suburbs in the neoliberal era. However, the relative youth of much of the area under study shelters it for the time being from the financial worries of aging.

Conclusion

The chapter has identified the determining features of the morphology of the North American dispersed suburb in an effort to better understand the dynamics of this urban form. It has adopted a backtracking approach, which relied on an observation of the morphology of the dispersed suburb to bring to light the laws responsible for the form taken by this type of development. The study has pointed to a predominant influence on this urban form of a generalized reliance on the automobile, to the extent that the morphology of dispersed suburbanism and full dependence on the car can be perceived as indissociable. The chapter maintains that the strength of this interconnection assures the reproducibility of this pattern of development by inhibiting the spawning of alternative suburban forms. The enduring nature of dispersed suburbanism is also a function of an enabling planning process and the existence of development-funding mechanisms adapted to the current neoliberal reality. But if these conditions sustain the production of new dispersed suburbanism environments, they also result in a tendency for these environments to deteriorate and filter down as they age.

REFERENCES

Appleyard, D. 1980. "Livable Streets: Protected Neighborhoods." *Annals of the American Academy of Political and Social Science* 451 (1): 106–17. https://doi.org/10.1177/000271628045100111.

Batty, M. 1999. "Editorial: A Research Program for Urban Morphology." *Environment & Planning B* 26 (4): 1–2. https://doi.org/10.1068/b260475.

Blais, P. 2011. *Perverse Cities: Hidden Subsidies, Wonky Policy, and Urban Sprawl*. Vancouver: UBC Press.

Bottles, S. 1987. *Los Angeles and the Automobile: The Making of a Modern City*. Berkeley: University of California Press.

Cervero, R. 2013. *Suburban Gridlock*. New Brunswick, NJ: Transaction.

Cervero, R., and K. Kockelman. 1997. "Travel Demand and the 3Ds: Density, Diversity, and Design." *Transportation Research Part D, Transport and Environment* 2 (3): 199–219. https://doi.org/10.1016/S1361-9209(97)00009-6.

Charmes, E. 2010. "Cul-de-sac, Superblocks and Environmental Areas as Supports of Residential Territorialisation." *Urban Design* 15 (3): 357–74. https://doi.org/10.1080/13574809.2010.487811.

Conzen, M.R.G. 2004. *Thinking about Urban Form: Papers on Urban Morphology, 1932–1998*. Oxford: Peter Lang.

Dear, M., and S. Flusty. 1998. "Postmodern Urbanism." *Annals of the Association of American Geographers* 88 (1): 50–72. https://doi.org/10.1111/1467-8306.00084.

Derrida, J. 1974. *Of Grammatology*. Baltimore: Johns Hopkins University Press.

Filion, P. 2007. *The Urban Growth Centres Strategy in the Greater Golden Horseshoe: Lessons from Downtowns, Nodes, and Corridors*. Neptis Studies on the Toronto Metropolitan Region. Toronto: Neptis Foundation.

Filion, P., T. Bunting, and K. Warriner. 1999. "The Entrenchment of Urban Dispersion: Residential Location Patterns and Preferences in the Dispersed City." *Urban Studies* 36 (8): 1317–47. https://doi.org/10.1080/0042098993015.

Filion, P., T. Bunting, K. McSpurren, and A. Tse. 2004. "Canada-U.S. Metropolitan Density Patterns: Zonal Convergence and Divergence." *Urban Geography* 25 (1): 42–65. https://doi.org/10.2747/0272-3638.25.1.42.

Filion, P., T. Bunting, D. Pavlic, and P. Langlois. 2010. "Intensification and Sprawl: Residential Density Trajectory in Canada's Largest Metropolitan Regions." *Urban Geography* 31 (4): 541–69. https://doi.org/10.2747/0272-3638.31.4.541.

Grant, J.L., and K. Perrott. 2011. "Where Is the Café? The Challenge of Making Retail Uses Viable in Mixed Use Suburban Developments." *Urban Studies* 48 (1): 177–95. https://doi.org/10.1177/0042098009360232.

Groth, P.E. 1981. "Streetgrids as Frameworks for Urban Variety." *Harvard Architectural Review* 2: 68–75.

Hackworth, J. 2006. *The Neoliberal City: Governance, Ideology, and Development in American Urbanism*. Ithaca, NY: Cornell University Press.

Harvey, D. 2005. *A Brief History of Neoliberalism*. Oxford: Oxford University Press.

Hulchanski, D. 2010. *The Three Cities within Toronto: Income Polarization among Toronto's Neighbourhoods*. Toronto: University of Toronto, Cities Centre Press.

Keil, R., and D. Young. 2011. "In-between Canada: The Emergence of the New Urban Middle." In *In-between Infrastructure: Urban Connectivity in an Age of Vulnerability*, ed. D. Young, P. Wood and R. Keil, 1–18. Kelowna, BC: Praxis Press.

Koolhaas, R. 1978. *Delirious New York: A Retroactive Manifesto for Manhattan*. New York: Thames & Hudson.

Kunstler, J. 1993. *The Geography of Nowhere: The Rise and Decline of America's Man-made Landscape*. New York: Simon & Schuster.

Lang, R.E. 2003. *Edgeless Cities: Exploring the Elusive Metropolis*. Washington, DC: Brookings Institution Press.

Langlois, P. 2010. "Municipal Visions, Market Realities: Does Planning Guide Residential Development?" *Environment & Planning B* 37 (3): 449–62. https://doi.org/10.1068/b34103.

Lehrer, U.A. 1994. "Images of the Periphery: The Architecture of FlexSpace in Switzerland." *Environment and Planning. D, Society & Space* 12 (2): 187–205. https://doi.org/10.1068/d120187.

Levy, A. 1999. "Urban Morphology and the Problem of the Modern Urban Fabric: Some Questions for Research." *Urban Morphology* 3: 79–85.

Lucy, W., and D.L. Phillips. 2000. *Confronting Suburban Decline: Strategic Planning for Metropolitan Renewal*. Washington, DC: Island Press.

Marshall, A. 2000. *How Cities Work: Suburbs, Sprawl, and the Roads Not Taken*. Austin: University of Texas Press.

Marglin, S.A., and J.B. Schor. 1990. *The Golden Age of Capitalism: Reinterpreting the Postwar Experience*. Oxford: Oxford University Press.

Moudon, A.V. 1997. "Urban Morphology as an Emerging Interdisciplinary Field." *Urban Morphology* 1: 3–10.

Moulaert, F., E. Swyngedouw, and P. Wilson. 1988. "Spatial Responses to Fordist and post-Fordist Accumulation and Regulation." *Papers/Regional Science Association. Regional Science Association. Meeting* 64 (1): 11–23. https://doi.org/10.1111/j.1435-5597.1988.tb01111.x.

Muller, P.O. 2004. "Transportation and Urban Form: Stages in the Spatial Evolution of the American Metropolis." In *New York: The Geography of Urban Transportation*, ed. S. Hanson and G. Giuiliano, 59–85. New York: Guildford Press.

Rowe, P.G. 1991. *Making a Middle Landscape*. Cambridge, MA: MIT Press.

Scheer, B. 2010. *The Evolution of Urban Form: Typology for Planners and Architects*. Chicago: Planners Press.
Schmid, C. 2012. "Patterns and Pathways of Global Urbanization: Towards Comparative Analysis." In *Globalization of Urbanity*, ed. J. Acebillo, J. Lévy, and C. Schmid, 51–78. Mendrisio, Switzerland: Accademia di architettura – Università della Svizzera Italiana.
Sewell, J. 1993. *The Shape of the City: Toronto Struggles with Modern Planning*. Toronto: University of Toronto Press.
Sieverts, T. 2003. *Cities without Cities: An Interpretation of the Zwishenstadt*. London: Spon Press. https://doi.org/10.4324/9780203380581.
Southworth, M., and E. Ben-Joseph. 1995. "Street Standards and the Shaping of Suburbia." *Journal of the American Planning Association* 61 (1): 65–81. https://doi.org/10.1080/01944369508975620.
Southworth, M., and E. Ben-Joseph. 2003. *Streets and the Shaping of Towns and Cities*. Washington, DC: Island Press.
Southworth, M., and P.M. Owens. 1993. "The Evolving Metropolis: Studies of Community, Neighborhood, and Street Form at the Urban Edge." *Journal of the American Planning Association* 59 (3): 271–87. https://doi.org/10.1080/01944369308975880.
TTS (Transportation Tomorrow Survey). 2006 Transportation Survey. Computer data retrieval. http://dmg.utoronto.ca/transportation-tomorrow-survey/tts-demographic-and-travel-summaries.
Vance, J.E., Jr. 1990. *The Continuing City: Urban Morphology in Western Civilization*. Baltimore: Johns Hopkins University Press.
Venturi, R., D. Scott Brown, and S. Izenour. 1977. *Learning from Las Vegas: The Forgotten Symbolism of Architectural Form*. Rev. ed. Cambridge, MA: MIT Press.
Walks, R.A. 2013. "Mapping the Urban Debtscape: The Geography of Household Debt in Canadian Cities." *Urban Geography* 34 (2): 153–87. https://doi.org/10.1080/02723638.2013.778647.
Wang, S., and P.J. Smith. 1997. "In Quest of a 'Forgiving' Environment: Residential Planning and Pedestrian Safety in Edmonton, Canada." *Planning Perspectives* 12 (2): 225–50. https://doi.org/10.1080/026654397364735.
Whitehand, J.W.R., and C.M.R. Carr. 2001. *Twentieth-century Suburbs: A Morphological Approach*. London: Routledge.

chapter six

The Paradox of Informality and Formality: China's Suburban Land Development and Planning

FULONG WU AND ZHIGANG LI

As a developing country with a history of a centrally planned economy, China has seen a fundamental dualism between its urban and rural areas. This dualism is an essential feature of its society and governance. Within this dualism the city is seen as the more modern industrialized part, under the management of the state, while rural areas have a more self-sufficient and underdeveloped economy. Land in urban areas is state owned, and urban households can own their "properties" only on the top of state land. Land tenure in rural areas is officially under collective ownership, but, in reality, farmers possess individual house plots as de facto private land, while their farmland is in the hands of village cadres. This rural-urban division poses a difficulty in placing the "suburban" in the governance system. Suburban land is a complex territory in transition from rural to urban systems. In this peri-urban area farmers' housing is, de facto, privately owned, and land tenure may comprise a mix of state-owned and collectively owned farmland.

Following the urban-rural dualism, suburban land development can have two different types: the more formal state-sanctioned land development and the more informal approach of farmers' self-building. The sources of development finance are thus very different. For formal development, the process starts from the acquisition of farmland or sometimes village land. The local government often sets up a corporation, which is responsible for the land development. The development corporation is also known as the "local investment platform" (see discussion below) because it uses the land acquired as collateral when applying to banks to obtain finance in infrastructure investment. After the completion of infrastructure development the land is then sold to developers through land auctions. The local government gets the revenue from land sales, while the development corporation receives the profit from the building of infrastructure. To some extent, the corporation is an investment arm of the local government. However, for the self-built informal development finance is usually taken from the famers' private savings or borrowed through family and social networks. The village may also use a "shareholding company" to mobilize capital to manage or invest in village assets. In some cities in southern China, such as Dongguan, the village shareholding company takes responsibility for providing public services as well as social welfare, which creates a heavy financial burden for the rural collectives. Some rural villages were near the brink of bankruptcy after the global financial crisis in 2008 because the downturn affected export-oriented manufacturing, and rental incomes from factory buildings and the housing of migrant workers declined.

In the last several decades China has experienced rapid suburban development. There are different explanations for such rapid change, including industrial growth in the suburbs (Zhu 2004), the role of local "state entrepreneurialism" (Wu and Phelps 2011), the demand of the new middle class for more exclusive and private "gated communities" (Zhang 2010), and the inflow of migrants into the suburbs through rural to urban migration (Wang, Wang, and Wu 2009). Although many factors can be attributed to the process of suburbanization, populations moved to suburbs for practical reasons. First, the central areas of Chinese cities have a high density and are quite congested. For a long time living space per inhabitant has been small. To improve living quality and enlarge the space of living, development had to be dispersed into the suburbs because it is simply too difficult to find available space inside the central areas. Second, Chinese cities faced economic restructuring. The central areas have been transformed from industrial and residential uses to office and commercial uses. As a result of land-use changes, population densities decreased in the central areas. Former "urban villages," which were created in a peri-urban location and have now effectively become inner suburbs, face high redevelopment pressures. Some villages have been demolished and converted into office use (Wu, Zhang, and Webster 2013a). Land-use changes in central areas and inner suburbs and the extension of suburbs through the outward movement of populations have together become a driving force for suburban land development.

This chapter focuses on the suburban land question (see chapters 1, 11, and 13 in this volume), especially land tenure, land management, and urban planning processes. The main research questions are how this form of land governance plays a role in Chinese suburban growth and what urban form and spatial pattern emerge from such a land development process. In this study we emphasize the form of land governance in creating Chinese suburban spatial forms in addition to the "flexible regime of accumulation" that created a "FlexSpace" in suburban Switzerland (Lehrer 1994), or social cultural dynamics that transformed American "edge cities" (Knox 2008). In the Chinese context land is the focus of the politics of suburban development. We conclude by commenting on the paradox of informality and formality and how they coexist as the foundation for post-reform suburban development in China.

The Institution of Growth: The Suburban Growth Machine

Land tenure is perhaps the most important institution that affects the pattern of land development. The property rights perspective tries to explain why modern capitalism is developed on the basis of defined property rights that can be transacted (de Soto 2000). However, whether formalizing land titles can generate economic growth is debatable (Gilbert 2002, 2007), as in developing countries there are traditional rights that are not necessarily in a legally recognized form. There are extensive studies on the Chinese land system and its implication for urban and suburban development (e.g., Hsing 2010; Lin and Yi 2011; Wang, Wang, and Wu 2009; Wu 2009; Wu, Zhang, and Webster 2013a; Zhu 2004). The unique Chinese rural and urban land dualism gives the local state great power to acquire the rural collective land to finance suburban land development.

How the local/municipal government is financed may affect the behaviour of the government in dealing with suburban land development. The locally defined and territorially bounded interests in land, with sufficient power to extract the benefit from local development, may lead to an entrepreneurial and growth-machine type of governance, typical of the United States (Logan and Molotch 1987), while in the UK and Canada the fiscal system is more centralized, and local entrepreneurial governance may be driven by the incentive devised by the central government to allocate the budget on a competitive base.

The local government in general strives to keep local taxes low in order to attract investment while depending on the transfer of central/upper-level funds to cover government spending on public services (see chapters 1, 11, and 13 in this volume). China, however, is quite unique in its local government finance.

China operates a centralized system where the central government and local governments have separate tax systems (known as *fen shui zhi*). Established in 1994 to cope with the declining central tax income after economic devolution, the central government set up its own tax base, leaving the local government 25 per cent of value-added tax. But the local government receives business tax. Along with economic devolution, the central government managed to download social expenditure and public service expenditure to the local governments, creating a fiscal deficit for local governments. However, the local government is able to develop land to increase its revenue from land sales. Land development profits became the major driver for suburban land development. Because there is virtually no property tax (except some limited fees for infrastructure construction), the local government is less interested in serving existing residents and more interested in new land requisition and the transfer of its use from agricultural to commercial and residential uses. In addition, property development can generate business tax from service sectors. However, in order to maximize income from commercial and residential land sales the local government needs to expand local economic activities. The local government tries to provide the land for industrial development at a lower price so as to enlarge the economic base.

The institution of public finance provides a strong incentive for the local government to develop and sell rural land. Owing to the unique urban-rural land-tenure system that grants the local government much stronger power in acquiring the land from rural areas, suburban land development becomes the most effective way to generate land revenue. In other words, Chinese municipalities control the entire rights of development of rural land within their jurisdiction boundaries. The farmers own their land collectively, which means they cannot sell the land directly to the users but have to transfer ownership to the municipal government. The municipality has the power of compulsory purchase to obtain rural land for urban development. But in order to conduct land development in suburbs, which is usually very costly, there has to be a third component in place: development finance. Without state support in development finance, large-scale suburban development is unlikely to occur based purely on commercial terms.

An institution innovation in development finance is the creation of state-backed "investment and capital mobilizing platforms" (*touzi rongzi pingtai*), which are state-owned enterprises formed to borrow money from the bank. In order to secure loans the investment platforms have to possess some assets. In fact, the most effective investment platforms are land corporations, which own the land. Pioneered in the development of the new suburban district of Pudong in Shanghai, a model of "virtual capital circulation" was created. The land corporation had no capital to develop suburban land but, using the land as collateral, it managed to draw capital from the bank system. After land development, the profit was used to pay back the loan, and the land corporation managed to complete the circulation of capital. This meant that the government could start up development without initial capital. Because the land corporations are backed by the local government and based on the credit of that government, the liability of "commercial" suburban land development is thus transferred to the government. It is estimated that local government debt could amount to 20,000 billion Yuan. The fluctuation of property markets could seriously jeopardize local government finance. Figure 6.1 shows the components of the Chinese suburban growth machine, which lays down the basic foundation of the suburban land question in China.

Processes of Suburban Development

Having identified the institutional set-up that forges the Chinese suburban growth machine in the previous section, in this section we will follow the theoretical framework proposed in Harris and Lehrer (chapter 1 in this volume) to examine the processes of suburban development. In particular, we examine three related issues: the degree of informality, the source of suburban residents, and the drivers of economic restructuring. First, informality, in the context of development, refers to informal development that does not conform to government regulation (Roy 2005, 2009). This form of irregular growth is widespread, not just as squatter development, but also as middle-class private housing areas where privatization and exception of government regulation allow for the contravention of formal regulations. In short, according to Roy (2005), informality is now a mode of metropolitan urbanization. It happens in different contexts, for example, as mafia

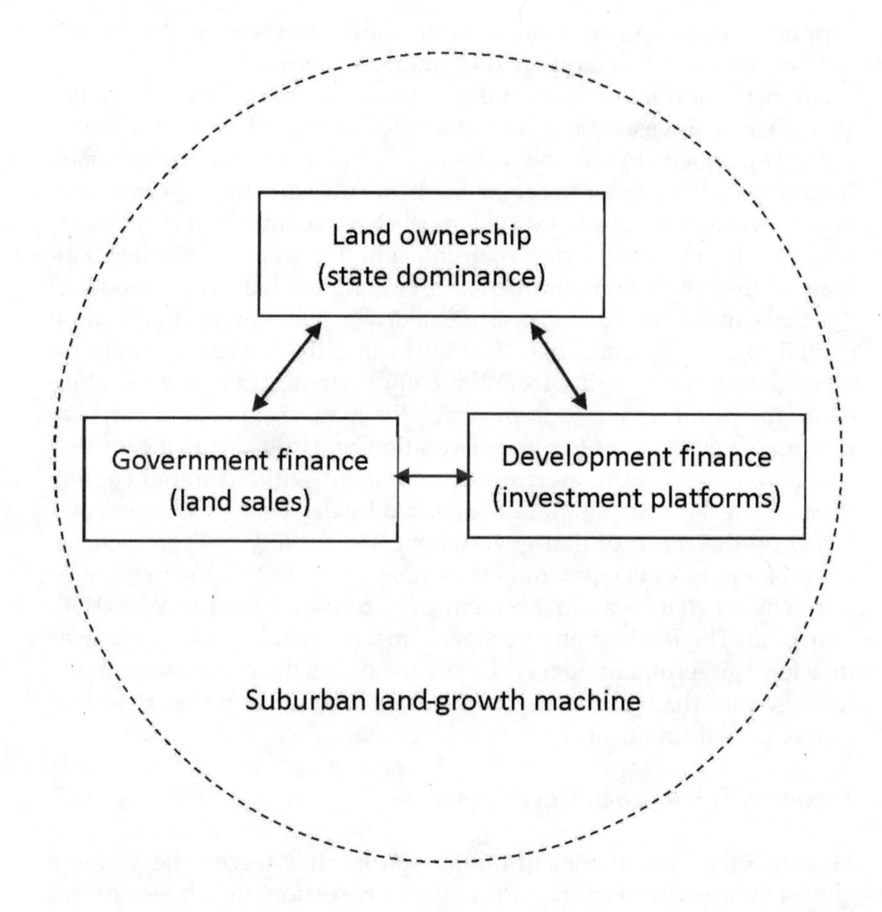

Figure 6.1 The Chinese suburban land-growth machine. Diagram: Fulong Wu

developers in India (Weinstein 2008) or pirate development and subdivision in Latin America (Doebele 1987; Gilbert 2007). In peri-urban areas Chinese urban villages exhibit four sources of informality: the dualistic and fragmented land ownership, lax land management and development control, informal service provision and management, and marginal and ambiguous status of village governance (Wu, Zhang, and Webster 2013a). The sources of informality are derived from the fact that rural land is largely outside the state's domain and hence has not been regulated in the same way as the state lands inside urban areas.

The process of suburbanization is a way to convert the rural land outside the state domain into a manageable state asset that is sold in the land market. Large-scale suburban land development in Chinese cities is highly regulated, follows a blueprint, and is often developed as a packaged mega project. Therefore, new suburban developments, especially industrial development through industrial parks and residential developments targeted at the middle class, take the form of formal development. On the other hand, the residual area that the formal land development could not cover (e.g., small land parcels, places near railway lines and under electric corridors), are left to informal development. Sometimes the pragmatic practice of land requisition contributes to the source of informal development. For example, to reduce the cost of monetary compensation local governments such as Guangzhou or Foshan allow some agricultural land to be developed into assets under village control. This collective ownership is subject to little formal development control and hence strengthens informality. The informal land development in Chinese suburbs is complementary to more formal, state-sanctioned land development.

Second, in terms of sources of suburban residents, Chinese suburbs see a mix of processes of urbanization and suburbanization. In addition to the outward movement of residents from central areas, the suburbs are also workplaces for rural migrants. The proportions of these two groups vary according to different locations and distance to the city. The development of gated estates and new towns also attracts people from smaller rural villages and towns in the suburb, as in Songjiang New Town in Shanghai (Shen and Wu 2013). In terms of the demand-side explanation, the suburb is not just a home for the upwardly mobile middle class to experience a dream of the good life, but also a workplace for rural migrants from the countryside to earn a living in the city. There are three main groups of suburban residents according to their origins: migrants from other places, residents relocated from the central city, and local natives living in either rural villages or small towns. While detailed data are not available to suggest their proportions, the relocated residents are dominant in housing near the metro station, while in the old town of Songjiang local natives are the main social group. For homebuyers, exclusive residential land use, green space, and high-quality amenities are important factors (Pow 2009; Wu 2010; Shen and Wu 2013). In addition to the quality of housing and living environment, investment in property is another motivation. Faced with a buoyant housing market and inflationary pressures since 2000, middle-class households have found investment in property to be a major way to preserve their wealth. Because of different demands the suburbs see the contrast between villas and apartments and between white-collar suburb, migrant suburb, and suburban new town. For white-collar suburb, typically built to a higher standard and a lower-density style, residents such as local civil servants, who are in a more advantageous position in China, are driven by the quality-of-life choice. For migrant suburbs, the attraction is due to cheaper rent and access to their workplaces. For suburban new towns with high-rise apartment buildings, in addition to older residents relocated by infrastructure and urban redevelopment in central areas, younger professionals choose the location because of cheaper housing.

Third, suburban land development is also driven by economic restructuring and the relocation of industries from central to periphery locations. In addition to a demand-side explanation, the supply-side explanation helps in understanding the role of the local government in developing industrial parks and extracting land-development profits from suburban expansion. China's economic development has been driven by fast industrialization. The growth of manufacturing industries, especially under foreign investment for the production of global economies, has increased the demand for industrial space. Because the central area is too expensive and lacks large land parcels, the new economy is exclusively enlarged through development zones and new towns; for example, the suburban district of Jiading in Shanghai is built around automobile industries and the creation of an "international automobile town." The Songjiang district of Shanghai has attracted universities and electronic manufacturing industries. The development of suburban industries created job sub-centres and led to the decentralization of the economy. This development is partially supported by the government's policy of restructuring the economy of large cities in China, for example, for Shanghai to build itself into the international economic, financial, trading, and shipping centres (the so-called four centres of Shanghai, which strives to become a global city).

Patterns of Suburban Space

In terms of suburban development patterns, Chinese suburbs demonstrate three salient features: the spread of suburban high-rise buildings; a high degree of heterogeneity in terms of ownership (owner-occupied housing versus "private rentals" – farmers' housing rented to migrants) and building forms (e.g., the villa vs. apartments); and significant industrial development. First, the suburban high-rise buildings are the usual form, and single-family housing is limited. In contrast to the standard single-family suburban sprawl in North America, the most common form of suburban development in China is the "new town." These new towns are planned suburban residential areas, often combined with large-scale industrial development and development zones. The absence of single-family detached houses is perhaps because it is difficult to build exclusively car-dependent suburbs. China has recently seen rising car ownership, but increasing automobile mobility is synchronized with the development of suburbs rather than as a pre-condition for suburban development. The growth of suburbs in this context is almost inevitably dependent upon the support of the government to build efficient roads and public transit to the suburbs. Large-scale informal suburban development is rare. Suburban land developments are combined with large-scale urban projects supported by the government as part of the overall structural transformation of urban areas. For example, another common type is the building of "university towns," a type of new town built for students and employees of universities relocated from the central areas. Because of the expansion of student recruitment in the 2000s Chinese universities required additional space. In the mid-2000s the land needed for universities was treated as "educational uses," which could be released more easily and at a lower price. This gave the government an opportunity to attract universities into the suburbs. As a consequence, large suburban new towns built around a university campus have been developed. These university towns are wholly planned, usually with shared public space and facilities for several universities in addition to their own campuses. In Panyu District of Guangzhou, for instance, so-called Guangzhou University Town is a typical cluster of ten universities and 300,000 students with a total land area of 40 km^2.

Second, Chinese suburban space is strongly bifurcated by different building forms and ownerships. Different building forms are obvious, but even for the same style of building, there might be different ownership types. For example, some rural villages in Beijing built high-rise residential buildings that are indistinguishable in appearance from other commodity housing estates (Figure 6.2). The only difference between housing in the formal market and village-built housing is their different ownerships. Housing built by farmers or their collectives belongs to "partial property right housing" (*xiao chanquan fang*).

Figure 6.2 "Informal" village housing built for an unauthorized housing market in Beijing, China. Photo: Fulong Wu

The notion of "partial" refers to the fact that the land is still legally under the rural collectives and has not been converted into official urban deeds. The Chinese suburbs see a wide spectrum from formal to informal housing. The partial property right housing is built for sale. But the housing in "urban villages" – rural villages encroached on by urban expansion (Wu, Zhang, and Webster 2013b) – is built mainly as informal "private rentals" to receive rural migrants. This is a result of fast urbanization and an influx of migrant workers into cities. The housing of new migrants is not considered in the government housing provision system. Neither can rural migrants afford the "commodity housing" built formally through land sales. The housing built by farmers in "urban villages" fills the gap between public and commodity housing and suits the preference of migrant workers who wish to minimize their costs (Figure 6.3). The quality of village rental housing is generally lower than standard commodity housing, but it evolves as the demand for facilities increases. For example, many apartment buildings are now equipped with an internet connection.

Figure 6.3 An urban village (Zhu Village) in Guangzhou, China.
Photo: Fulong Wu

Given the normal standard of high density in Chinese cities, lower-density detached housing enclaves in the suburbs are exclusively for the rich. Some choose them as a lifestyle preference, but more use the house as second or holiday homes. In terms of the lifestyle preference, the new rich pursue these places in the context of a rising awareness of the aesthetic quality of the built environment (Pow 2009; Wu 2010; Shen and Wu 2012). To suit their preferences, real estate developers actively promote an imagined Western life as a marketing tactic. Some projects use Chinese traditional styles, but such structures are not common. Some even create a residence wholly engineered to replicate Western style. For example, Thames Town of Shanghai was developed using a model of an English market town (Shen and Wu 2012). The contrast is sharp between highly functional village housing and heavily decorated dwellings (often in Western styles) in gated communities.

Third, the Chinese suburbs see a significant proportion of industrial land uses. In fact, as discussed earlier, large-scale suburbanization in China has been supported by the government and led by industrial development. Large-scale residential sprawl without accompanying industrial decentralization was absent in China. In a sense, contemporary suburban land development in China is similar to "post-suburbia" (Phelps and Wu 2011), which refers to the phase after residential suburbanization. Post-suburbia is characterized by the creation of a suburban economy and the retrofit of low-density residential suburbs. However, in North America the growth of a suburban economy is largely associated with post-industrial development. The concepts of edge city (Garreau 1991), FlexSpace (Lehrer 1994), and metroburbia (Knox 2008) are seen as the development of office space outside the downtown. In contrast, Chinese cities are still in the stage of industrialization, and suburban land development is heavily driven by the demand for manufacturing industries. As China is becoming the "workshop of the world," its suburban space is a product of both globalization and local development. The result is that Chinese suburban land development is characterized by the emergence of industrial development zones. For production services there is still a strong centripetal tendency in the city of Guangzhou (Yi, Yang, and Yeh 2011). However, business parks may appear as a specific tactic of government-sponsored local development. For example, in the city of Kunshan near Shanghai, a business park (Huaqiao Business Park) has been deliberately created as an edge city of Shanghai to capture back-door office jobs, although like Tysons Corner outside Washington, DC, the edge city of Huaqiao is outside the jurisdiction of Shanghai (Wu and Phelps 2008). The slogan of Huaqiao Business Park is to develop a "business satellite town for globalizing Shanghai." The economy is oriented towards the following four pillar industries: the regional headquarters of manufacturing industries, back-office and data processing of financial organizations, information technology outsourcing and business process outsourcing, and logistics and purchase centres. Other examples include the development of the Beijing Economic and Technological Zone in the new town of Yizhuang on the outskirts of Beijing, where an office park has been built as part of an overall development strategy, especially for small- and medium-size enterprises that demand cheaper space but within purpose-built buildings (Wu and Phelps 2011), and Zhangjiang High-tech Park in Shanghai Pudong with a cluster of biotech and integrated circuits R&D (Zhang and Wu 2012).

Because manufacturing industries drive the development of suburban economies, for the development zones or industrial parks there is an issue of mismatch between jobs and housing. The suburban industrial parks see the concentration of jobs, but there is insufficient affordable housing for industrial workers. Some have to commute from the central area and other suburban areas. More jobs than homes are available, leading to an increasing separation between workplaces and homes. As a result, city planners are asked to "balance job and housing" (*zhizhu pingheng*) or to "integrate industries into the city" (*chancheng ronghe*) – literally, to identify more land for residential uses.

Planning the Chinese Suburbs

What is the contribution of China's planning system to its suburban land-development processes? Does the land-use planning system effectively manage and control the speed of urban expansion? Any answers must consider different scales of governance. At the national level there is no specific urban policy that affects the process of suburbanization. The most relevant policy is the protection of agricultural land by the central government. Concerning food security, the central government has set up a redline policy of 1.8 billion mu (equivalent to 108 million hectares) "basic agricultural land," which forbids the conversion of the designated basic agricultural land into non-agricultural or urban uses. In addition, to control the speed of land development, the policy is implemented through land-development quotas, which are issued annually by the central government (Ministry of Land and Resources) to local governments. In theory, the policy might constrain the speed of urban expansion as the suburban land surrounding the central city is usually used intensively as agricultural land for vegetable production. But the policy has some flexibility. That is, the protection of agricultural land is enforced with the quantity, not the actual uses, of particular land parcels. In this sense the policy of designated basic agricultural land is implemented through "indicator management," which is different from the practice of "land-use planning" in other countries. This policy then gives the local government a chance to manipulate land use by developing the most accessible and profitable land near the city and balancing the quota by various efforts such as converting some non-agricultural land in more remote areas to agricultural land. As a result, while the national policy may affect the overall speed of land development, it could not effectively regulate the expansion of urban land in suburban areas. The local government thus has some discretion on how suburban land is developed.

The most significant national policy that affects suburban land development is the tax-sharing system, as discussed earlier, which allows the central government to collect the most taxes and decentralize some expenditure to the local government. The fiscal gap is then left to local governments to fill from their "entrepreneurial" engagement to sell state land to gain the revenue from land development. Land revenue becomes an important source of income for local governments (Tao et al. 2010). Therefore, local governments have strong incentives to promote suburban land development. At the local level the planning system plays an important role in encouraging rather than controlling urban expansion. From strategic urban plans to urban master plans the system holds an expansionist approach in suburban development. Strategic urban plans often propose some ambitious new development by setting up a development zone, creating a new growth corridor, or annexing rural counties into new urban districts so as to open the space of growth, while the urban master plan is used to prepare the land use of new towns and urban design creates a more specific image of new town. At the city-region level the urban system plans often encourage the development of vast rural areas between two nearby large cities. For example, Xi'an and Xianyang have proposed a new development zone called Xi-Xian New Area. Guangzhou and Fuoshan have formulated a strategy to link up through intercity integration (*tong cheng hua*). In the east of Zhengzhou a brand-new area has been developed called Zhengzhou East District. It was designed by the renowned Japanese architect Kisho Kurokawa, who master-planned a water landscape by creating an artificial lake surrounded by a ring of skyscrapers. In suburban Shanghai, Songjiang New Town is built with relocated university campuses forming a university-town. To add to its attractiveness, Songjiang New Town asked a British planning firm, Atkins, to create an English market town called Thames Town (Shen and Wu 2012). Land-use planning in the context of rapidly growing Chinese suburbs is used as a tool to promote suburban land development. Place marketing rather than development control is the major function of planning (Wu 2007). Place marketing includes, for example, the creation of an artificial and attractive place name. Planners are asked to borrow examples from new town development in Singapore or the UK to create a brand for otherwise unknown agricultural fields in the suburbs.

In this sense, although suburban land development in China uses the market approach (e.g., the involvement of land developers), the process of development is not entirely informal but rather initiated by local planning policies. It is interesting to note that this approach is not a "deregulated" neoliberal approach, because planning still plays a rather active role in suburban land development, similar to the overall development approach in China. In many aspects the status of planning has been strengthened rather than reduced. But the role of protecting agricultural land and rural communities is rather weak, because the planning system in China is always state led and top down. Since economic devolution local governments have begun to dominate the process of land-use planning and turn it into a device of place promotion (Wu 2007). The weakness in development control is due to the marginal position of neighbourhoods or farmers' villages in this process of land development. In other words, residents are not able to influence the planning system to resist land development imposed by municipalities. Confronted with the rapid pace of development and incapable of directing the future course of development, local farmers themselves try to develop the land in various informal ways – for example, self-building the land into private rentals or leasing the land to small developers – when opportunities arise to benefit from development. This has resulted in a patchy pattern of development and a juxtaposition of contrasting landscapes of gated communities and the urban villages lived in by rural migrants.

As discussed earlier, China's urban and rural areas are subject to different land-use managements, although the Urban-Rural Planning Act of 2008 tried to extend urban land planning to rural areas. In general, traditional rural land management is laxer, and the monitoring of development and enforcement of development control is understandably less effective. The weak position of rural land management and the power of farmers in rural areas make it easier to develop land in rural areas rather than in urban brownfields. Master-planned new towns implemented by the local governments are a favourable approach to land development. Very often these new towns are proposed as a compact form of development, similar to "new urbanism" in the United States (Katz 1994). The density of Chinese suburbs is generally higher than that of their counterparts in the West. It is not unusual that new development can comprise residential buildings of up to thirty floors (Figure 6.4). The suburban land development of homogeneous single-family houses is rare or constitutes only a small proportion of larger developments. For example, in Shanghai's Songjiang New Town, with a population of 800,000, only a small area of 1 km^2 of Thames Town has a low-density form of detached houses or so-called villas.

Figure 6.4 Suburban housing in Jiading District of Shanghai, China. Behind the villa-style housing are the "ordinary" suburban high-rises widely seen in Chinese suburbs. Photo: Fulong Wu

Different Perspectives on the Paradox of Suburban Land Development

Suburban land development in China can been seen from three perspectives: suburbanization as the development and conversion of the way of life, suburbanization as an outlet for capital investment, and suburbanization as a process of economic and technological development. This section will discuss these perspectives and how applicable they are to the Chinese context.

First, suburbanization can be seen as the development of a distinctive way of life between the city and countryside. The concept of "Garden City" deploys the combination of urban and rural as an advantage of smaller communities in the peripheral urban areas. The classic concentric urban model features the differentiation of lifestyles along with the changing distance to the city centre. Urban land economics explains this variation by noting the different preferences for land value and transport (commuting) costs. While the inner urban area is becoming urban, the peri-urban area, juxtaposed with rural and urban uses, is becoming suburban in the United States (Hanlon 2010). At the same time, the suburban life is accompanied by lower-density, single-family housing and gendered division of labour and, to a greater extent, separation of family life and workplaces (Beauregard 2006). Is this lifestyle explanation applicable to Chinese cities? The Chinese suburb attracts a diverse population: rural migrants to the cities, local residents originally living nearby in the rural area or small towns, and residents relocated from the central city (Shen and Wu 2013). While some inner-urban residents are relocated by government redevelopment programs and hence are not driven by lifestyle choice, the upwardly mobile middle class has become more conscious of the suburban living environment – especially the quality of the built environment in lower-density villa areas. In gated communities security and landscaping quality are attractive features (Zhang 2010; Pow 2009; Wu 2010). Those residents who want to move to the suburbs pay more attention to the quality of living environment and housing conditions. However, richer residents may not physically relocate to the suburbs but just "consume" the suburban lifestyle on weekends and holidays, as many houses are bought as second or third properties. In contrast, the move to the suburbs of rural migrants is almost always determined by jobs: the development of industries in Chinese suburbs offers a source of low-paid, labour-intensive employment. Migrants live in rental apartments rather than owning homes, a situation that is different from Americans living in "ethnoburbs," where ethnic groups become homeowners (Price 2012).

Second, the suburban development could be perceived as an important "spatial fix" of capitalist development from the radical political-economic point of view. Linked to the seminal idea of capital switch and the circuits of capital (Harvey 1985), suburbanization is read as an important outlet for capital to cope with over-accumulation and the dilemma of socio-spatial development. The suburban land development in the United States has been supported by federal urban and economic policies to boost post-war housing consumption. This reading has not been adequately explored in the Chinese context, but it is sufficient to suggest that the development of suburbs plays an important role in facilitating the transformation of Chinese cities. As a consequence of buoyant housing markets Chinese cities are increasingly driven by investment in land development. The building of commodity housing and intensified suburban land development are associated with the strategy of the state to cope with the Asian financial crisis of 1997 and the global financial crisis in 2008. The two crises have forced the state to adopt more aggressive investment approaches to boost domestic consumption to compensate for the decline in exports. Suburban land and housing have been seen as an important investment outlet for sustaining capital circuits.

Third, suburbanization can be interpreted as the process of economic decentralization from the central location to edge cities (Garreau 1991), together with the technological changes that make suburban locations a possible place for innovation and business (Fishman 1987) and with the outward movement of populations. This view seems to fit Chinese suburban economic development, as the establishment of economic and technological zones in the suburbs has greatly transformed the landscape of Chinese cities. China's suburban expansion is largely driven by foreign direct investment in factories and export-oriented industries. The improvement of road infrastructure and intercity railways makes it possible to choose exurban locations for new industries. Recent emphasis on "indigenous innovation," the shift towards higher ecological-quality business parks, and the development of "eco-cities" have further facilitated the outward movement of economic activities. For example, Shanghai's Zhangjiang high-tech park is located in the new development area of Pudong, away from the more mature areas in the downtown (Zhang and Wu 2012). Further, the park has now been scaled up to become a nationally designed cluster of high-tech developments comprising one central zone and eighteen other parks across municipalities, which has stimulated Shanghai's suburban economies. Through suburbanization a polycentric urban structure has emerged in Chinese cities (Feng, Wang, and Zhou 2009). The relocation of universities from central areas to the outskirts and the development of new campuses in the suburbs have also been an important driving force in post-industrial suburban economic changes. The growth of industrial and services economies in the suburbs show that China's new round of suburbanization has a much more complex pattern than homogeneous residential dispersal. Terms such as FlexSpace (Lehrer 1994), metroburbia (Knox 2008), and post-suburbia (Phelps and Wu 2011) attempt to describe these new patterns of suburbanization in association with a metropolitan-wide spatial reconfiguration of work and residence.

Conclusion

Chinese suburban land development is largely state sanctioned. The state owns the monopolistic right in the acquisition of rural land and releases the land to the market. Other forms of informal suburban land development, for example, spontaneous conversion of farmers' land into private rentals in "urban villages," are not officially recognized. In the state-endorsed land market there are multiple forms of suburban development, including industrial parks, university towns, shopping malls, and large gated estates. The newly built estates under the name "commodity housing" attract residents from central areas. Some of them are relocated by urban redevelopment in the central area for commercial and office uses. Others are attracted by lower house prices in the suburbs. A few have even bought suburban houses for the quality of the environment. However, most villa owners use them as second or third homes, as investments, or for holidays. In contrast, in peri-urban areas urban village housing accommodates rural migrant workers in factory dormitories. In this sense Chinese suburban development is a result of both suburbanization and urbanization, where industries are developed and the working class is formed in addition to the relocated middle class.

The development of suburban land is a process of formalization of land tenure, although there are various forms of informality in Chinese peri-urban and rural areas (Wu, Zhang, and Webster 2013a). The urban rural dichotomy is a division between the state and non-state systems. The urban belongs to the industrialized state system, while the rural is a largely self-sufficient and self-dependent non-state system. The extension of the formal state system into the rural areas creates a juxtaposition of different forms of land tenure. The contrast between urban villages and commodity housing estates includes not only the differences in landscapes of irregularity and enclosure, but also their different forms of land tenure. While Harris, Lehrer, and Bloch (2013, 21) argue that "suburbs constitute one of the main places in which market capitalism is penetrating and expanding," the Chinese suburban land question indicates that the route of becoming (sub)urban is controlled by the state. Chinese suburbs will continue to reveal the complexity of land questions.

REFERENCES

Beauregard, R. 2006. *When America Became Suburban*. Minneapolis: University of Minnesota Press.

de Soto, H. 2000. *The Mystery of Capital: Why Capitalism Triumphs in the West and Fails Everywhere Else*. New York: Basic Books.

Doebele, W.A. 1987. "The Evolution of Concepts of Urban Land Tenure in Developing Countries." *Habitat International* 11 (1): 7–22. https://doi.org/10.1016/0197-3975(87)90030-0.

Feng, J., F. Wang, and Y. Zhou. 2009. "The Spatial Restructuring of Population in Metropolitan Beijing: Towards Polycentricity in the Post-reform Era." *Urban Geography* 30 (7): 779–802. https://doi.org/10.2747/0272-3638.30.7.779.

Fishman, R. 1987. *Bourgeois Utopias: The Rise and Fall of Suburbia*. New York: Basic Books.

Garreau, J. 1991. *Edge City: Life on the New Frontier*. New York: Doubleday.

Gilbert, A. 2002. "On the Mystery of Capital and the Myths of Hernando de Soto: What Difference does Legal Title Make?" *International Development Planning Review* 24 (1): 1–19. https://doi.org/10.3828/idpr.24.1.1.

Gilbert, A. 2007. "The Return of the Slum: Does Language Matter?" *International Journal of Urban and Regional Research* 31 (4): 697–713. https://doi.org/10.1111/j.1468-2427.2007.00754.x.

Hanlon, B. 2010. *Once the American Dream: Inner-ring Suburbs of the United States*. Philadelphia: Temple University Press.

Harris, R., U. Lehrer, and R. Bloch. 2013. "The Suburban Land Question." Discussion paper, Global Suburbanisms Project Workshop on Land. Montpellier, France, 21–23 October 2012. Rev. MS.

Harvey, D. 1985. *The Urban Experience*. Baltimore: Johns Hopkins University Press.

Hsing, Y.-T. 2010. *The Great Urban Transformation: Politics and Property in China*. New York: Oxford University Press. https://doi.org/10.1093/acprof:oso/9780199568048.001.0001.

Katz, P. 1994. *The New Urbanism: Toward an Architecture of Community*. New York: McGraw-Hill.

Knox, P. 2008. *Metroburbia*. New Brunswick, NJ: Rutgers University Press.

Lehrer, U. 1994. "Images of the Periphery: The Architecture of FlexSpace." *Environment and Planning. D, Society & Space* 12 (2): 187–205. https://doi.org/10.1068/d120187.

Lin, G.C.S., and F. Yi. 2011. "Urbanization of Capital or Capitalization on Urban Land? Land Development and Local Public Finance in Urbanizing China." *Urban Geography* 32 (1): 50–79. https://doi.org/10.2747/0272-3638.32.1.50.

Logan, J.R., and H. Molotch. 1987. *Urban Fortunes: the Political Economy of Place*. Berkeley: University of California Press.

Phelps, N., and F. Wu, eds. 2011. *International Perspectives on Suburbanization: A Post-suburban World?* Basingstoke, UK: Palgrave Macmillan. https://doi.org/10.1057/9780230308626.

Pow, C.P. 2009. *Gated Communities in China: Class, Privilege and the Moral Politics of the Good Life*. London, New York: Routledge.

Price, M. 2012. "Ethnoburb: The New Ethnic Community in Urban America." *Annals of the Association of American Geographers* 102 (1): 254–6. https://doi.org/10.1080/00045608.2011.624966.

Roy, A. 2005. "Urban Informality. Towards an Epistemology of Planning." *Journal of the American Planning Association* 71 (2): 147–58. https://doi.org/10.1080/01944360508976689.

Roy, A. 2009. "Why India Cannot Plan its Cities. Informality, Insurgence and the Idiom of Urbanization." *Planning Theory* 8 (1): 76–87. https://doi.org/10.1177/1473095208099299.

Shen, J., and F. Wu. 2012. "The Development of Master-planned Communities in Chinese Suburbs: A Case Study of Shanghai's Thames Town." *Urban Geography* 33 (2): 183–203. https://doi.org/10.2747/0272-3638.33.2.183.

Shen, J., and F. Wu. 2013. "Moving to the Suburbs: Demand-side Driving Forces of Suburban Growth in China." *Environment & Planning A* 45 (8): 1823–44. https://doi.org/10.1068/a45565.

Tao, R., F.B. Su, M.X. Liu, and G.Z. Cao. 2010. "Land Leasing and Local Public Finance in China's Regional Development: Evidence from Prefecture-level Cities." *Urban Studies (Edinburgh)* 47 (10): 2217–36. https://doi.org/10.1177/0042098009357961.

Wang, Y.P., Y. Wang, and J. Wu. 2009. "Urbanization and Informal development in China: Urban Villages in Shenzhen." *International Journal of Urban and Regional Research* 33 (4): 957–73. https://doi.org/10.1111/j.1468-2427.2009.00891.x.

Weinstein, L. 2008. "Mumbai's Development Mafias: Globalization, Organized Crime and Land Development." *International Journal of Urban and Regional Research* 32 (1): 22–39. https://doi.org/10.1111/j.1468-2427.2008.00766.x.

Wu, F. 2007. "Re-orientation of the City Plan: Strategic Planning and Design Competition in China." *Geoforum* 38 (2): 379–92. https://doi.org/10.1016/j.geoforum.2006.05.011.

Wu, F. 2009. "Land Development, Inequality and Urban Villages in China." *International Journal of Urban and Regional Research* 33 (4): 885–9. https://doi.org/10.1111/j.1468-2427.2009.00935.x.

Wu, F. 2010. "Gated and Packaged Suburbia: Packaging and Branding Chinese Suburban Residential Development." *Cities (London)* 27 (5): 385–96. https://doi.org/10.1016/j.cities.2010.06.003.

Wu, F., and N.A. Phelps. 2008. "From Suburbia to Post-suburbia in China? Aspects of the Transformation of the Beijing and Shanghai Global City Regions." *Built Environment* 34 (4): 464–81. https://doi.org/10.2148/benv.34.4.464.

Wu, F., and N.A. Phelps. 2011. "(Post) suburban Development and State Entrepreneurialism in Beijing's Outer Suburbs." *Environment & Planning A* 43 (2): 410–30. https://doi.org/10.1068/a43125.

Wu, F., F.Z. Zhang, and C. Webster. 2013a. "Informality and the Development and Demolition of Urban Villages in the Chinese Peri-urban Area." *Urban Studies (Edinburgh)* 50 (10): 1919–34. https://doi.org/10.1177/0042098012466600.

Wu, F., F.Z. Zhang, and C. Webster, eds. 2013b. *Rural Migrants in Urban China: Enclaves and Transient Urbanism*. London: Routledge.

Yi, H., F.F. Yang, and A.G.O. Yeh. 2011. "Intraurban Location of Producer Services in Guangzhou, China." *Environment & Planning A* 43 (1): 28–47. https://doi.org/10.1068/a42460.

Zhang, L. 2010. *In Search of Paradise: Middle Class Living in a Chinese Metropolis*. Ithaca, NY: Cornell University Press.

Zhang, F.Z., and F. Wu. 2012. "'Fostering Indigenous Innovation Capacities': The Development of Biotechnology in Shanghai's Zhangjiang High-tech Park." *Urban Geography* 33 (5): 728–55. https://doi.org/10.2747/0272-3638.33.5.728.

Zhu, J. 2004. "From Land Use Right to Land Development Right: Institutional Change in China's Urban Development." *Urban Studies (Edinburgh)* 41 (7): 1249–67. https://doi.org/10.1080/0042098042000214770.

chapter seven

Comparing Recent Suburban Developments in Austria and the Netherlands

WOLFGANG ANDEXLINGER, PIA KRONBERGER-NABIELEK,
AND KERSTEN NABIELEK

Since the beginning of the twentieth century, an ongoing process of suburbanization has widely disarranged traditional settlement structures in Europe. Former agricultural land and natural areas around villages and cities have been transformed into hybrid built-up areas combining rural and urban functions. In particular, the development of urban areas on the periphery has accelerated in response to growing welfare, global economic forces, improved transportation links, and increased personal mobility. Thus, structures have developed whose centres, peripheries, or edges are no longer clearly recognizable (Eisinger and Schneider 2005, 11). Work, recreation, shopping, and other activities are no longer tied to the place of residence. Instead, these functions are distributed over a wider area (Andexlinger et al. 2005, 49). Urban developments on the periphery of European villages and cities have resulted in suburban areas that are characterized by a large degree of spatial and functional heterogeneity (see Bryant, Russwurm, and McLellan 1982; Frijters et al. 2004; Gallent, Andersson, and Bianconi 2006). These areas are shaped by the spatial coexistence of large new developments and the leftovers of former agricultural, natural, and industrial functions. In this way new residential areas, recreational parks, commercial areas, office complexes, and retail centres have emerged next to long-established land uses. As well, large-scale developments can be found between small-scale structures, as well as urban functions beside rural functions. These components of peripheral development are generally the same across Europe, having been brought about by similar economic, demographic, and social forces. They include demographic growth, liberalization of the economy, rise of an affluent society marked by consumerism, and individual mobility.

Seen from an aerial point of view, these developments display a mixture of functions that are spatially separated from each other by clear borders, such as leftover green spaces, fences, hedges, traffic spaces, transport infrastructure, and water bodies, or by a distinctive architectural "themed" layout (P. Nabielek 2011).

Approach and Objectives

In the past decades much research has been devoted to changes in suburban neighbourhoods and the reasons for the emergence of such fragmented areas. We may cite, among many others: splintering urbanism (Graham and Marvin 2001); *zwischenstadt* (Sieverts 1999); *stadtland schweiz* (Eisinger and Schneider 2005); in-between infrastructure (Young, Burke Wood, and Keil 2011); shadowland (Hamers and Rutte 2008) and *tussenland* (Frijters et al. 2004); *cittá diffusa* (Secchi 1997); territories of new modernity (Viganò 2001); spread city (Webber 1998); and *annähernd perfekte peripherie* (Campi, Bucher, and Zardini 2000). What all these studies have in common is that they deal with the phenomenon of advancing urbanization, describing it as a structure that has taken shape through countless individual decisions.

As Neil Brenner (2014) points out, the process of urbanization cannot be understood as simply a result of population growth or as a replication of city-like settlements across the earth. Neither can it be explained by traditional notions of a hinterland that is being enclosed, operationalized, designed, and planned to support the continued agglomeration of capital, labour, and infrastructure. Instead, Brenner calls for a new understanding of urbanization: "A new understanding of urbanization is needed, which explicitly theorizes the evolving, mutually recursive relations between agglomeration processes and their operational landscapes, including the forms of land use intensification, logistical coordination, core-periphery polarization and socio-political struggle that accompany the latter at all spatial scales" (2014, 21).

From this point of view it seems important to ask about the role played by planning within the process of urbanization. Europe has always been heterogeneous in terms of its administrative and corporate cultures, economic systems, legislation, and planning systems. Furthermore, many European nations have imposed planning restrictions to prevent or control out-of-town developments (Evers 2004). In this context three general questions need to be discussed: How "generic" is the European suburb in reality? What kind of planning approaches have been developed in various European countries to respond to the local and global driving forces described above? And, finally, to what extent does spatial planning policy influence the appearance of European suburbs today? This chapter does not give comprehensive answers to all of these questions, but it considers specific developments in two European countries.

Comparing Suburban Developments in Austria and the Netherlands

Having stated the desired contribution of this chapter to current scientific debates about suburbanization in Europe in the previous section, the following sections will describe the results of a comparative study of recent suburban developments in Austria and the Netherlands. Both countries have been undergoing substantial suburbanization for decades. Other key features they have in common are high living standards and a severely limited settlement area dictated by geographical conditions. In Austria the settlement area is limited to one-third (BEV 2011) of the country's territory because of its extensive mountainous terrain. In the Netherlands the most densely populated areas of the country are located on drained marshlands that must be protected against flooding by a vast, complex system of embankments and groundwater pumps. The high investment costs linked to reclaiming and maintaining land have led to a cautious treatment of land use.

Whereas the geographical restrictions and the economic and sociopolitical conditions prevalent in the two countries seem to be comparable, the spatial planning cultures are strikingly different. In the Netherlands, since the 1950s, a systemic and integrated planning approach involving effective national planning policies has had a strong influence on spatial developments. In Austria, however, decisions about urban developments are predominantly taken at the federal state and communal level, which has led to a more organic, ad hoc type of spatial planning.

By comparing suburban developments in both countries, we wish to show to what extent the specific national planning cultures have influenced spatial developments in their urban fringes. To this end, the chapter will compare the spatial planning approaches and spatial developments in both countries. The spatial analysis is based on land-use studies and visual analysis. In order to provide an even more specific depiction, we will look at recent suburban developments in four regions (two per country): the Vienna region and North Tyrol in Austria (A); Rotterdam-The Hague and Flevoland in the Netherlands (NL). Whereas Vienna (A) and Rotterdam-The Hague (NL) are highly urbanized and have a metropolitan character, North Tyrol (A) and Flevoland (NL) are (formerly) rural areas with a more regional character. We will examine suburban developments after 1990 in terms of housing, recreation, retail, and commerce.

Planning Culture and Suburban Developments in Austria

Spatial Planning in Austria: Local and Ad Hoc

Austria is situated in the eastern Alps and its surface area totals 83,879 km^2. Some 60 per cent of this area has a mountainous character. Only 32 per cent lies below 500 metres above sea level and 43 per cent is woodland (Statistisches Handbuch Bundesland Tirol 2014). Owing to natural conditions and its topographic situation, only 37 per cent of Austrian ground surface is suitable for sustained settlement purposes (Tötzer, Loibl, and Steinnocher 2009).

The Austrian spatial planning tradition diverges completely from the Dutch one. In Austria, spatial planning is a rather young field of study: it was introduced as a branch of learning in the higher education landscape only in the 1970s.

Although spatial planning is viewed as an important topic in professional circles, until recently it has not been possible to speak of a proper planning tradition, above all because competencies in the field of land use and spatial planning span a diversity of institutions and policy areas. For this reason clear structures that are binding for all have not taken root. Indeed, competent jurisdictions within Austria are strongly dispersed: the nine provinces of Austria oversee supra-local spatial planning; while individual local authorities are responsible for local decisions. The federal state, for instance, is in charge of transport issues, but in many other policy areas that affect spatial development there is simply no high-level regulatory power.

Regarding the distribution of competencies related to land use, the prevailing axiom is that fundamental decisions should be made at the local level. In spite of the importance of spatial development, the federal government has no land-use planning role, except in terms of infrastructural tasks of national importance. Neither is there a federal regional planning act, which would lay down general requirements. Instead, a regional planning conference (ÖREK) has been established; this planning coordination mechanism at the national level is supported by the federal state, the provinces, and the local authorities. However, the ÖREK cannot make legally binding decisions – it can only make recommendations. Indeed, in Austria the provinces are in charge of land-use legislation and determine supra-local rules. Since no such decisions are made at the federal level, rules governing land use are different in each of the nine provinces, in regard to both land use and building laws. The provinces are also in charge of supra-local planning. Provincial development programs and regional development programs are prepared at this level and are binding for the local authorities involved. However, the real power of decision in the field of land use lies not at the provincial level, but at the local level. Within the framework of planning, decisions related to spatial development are made locally and directly concern local development. Decisions made at a local level have to be confirmed by the provincial government according to the provincial planning instrument concerned. In this respect the provincial government acts as a supervisory body.

At the local authority level the following instruments generally are available: the local spatial development plan, the land-use plan, and the zoning plan. The local council acts as the primary authority on matters of spatial planning, but in terms of building issues the mayor has primary authority, while the council has a secondary role.

Thus, in Austria the distribution of competencies is strongly anchored at the local level, that is, within local authorities. According to this sovereignty principle, decisions that shape the spatial development of a town ostensibly are made according to its internal point of view. Therefore, every town and village is viewed as an "island" and future developments are steered on the basis of isolated views. Supra-local land-use planning at the provincial level has few tools at its disposal that can control spatial development at a higher governance level: protected landscape areas, dangerous zone maps, and other plans (sectoral) may be mentioned here.

Although these planning regulations may have an influence on Austrian spatial development through their prohibitions and restrictions, they do not make it possible to deal with spatial issues through new strategies that would be authoritative for all concerned. In the past decades the absence of a binding, nation-wide, spatial development approach has meant that land consumption continues to be high across Austria as a whole.

Recent Suburban Development in Austria

Overall, Austria is experiencing population growth, but the country's persistently high level of land consumption cannot be blamed on this trend. As in most industrialized countries, land consumption has been decoupled from population growth for a long time. A rather modest population increase of +1.1 per cent (Statistik Austria 2013) contrasts starkly with an unrestrained 10 per cent growth of land for settlement and transport (BEV 2009). Every day, on average, 10 hectares are added for construction and transport purposes (Umweltbundesamt 2013a). When one adds the land surface claimed for other infrastructure (e.g., utility areas, waste disposal sites, storage yards), amounting to 12.5 hectares/day, in total 22.5 hectares are being given a new "use" (ibid.). This is the equivalent of 17.5 football fields every day.

Today about 17 per cent of the total ground surface suitable for settlement is being used (ibid.). According to Guiding Principle 13 in the Austrian Sustainability Strategy (BMLFUW 2002), the daily claim for additional land surface by construction and transport should have been reduced to 2.5 hectares/day by 2010. Until now, this target has been widely missed.

The sustained, very high level of land consumption in Austria is, among other factors, due to house building. Although between 1990 and 2011 the population grew by about 10 per cent (Statistik Austria 2013), the number of households went up by 25 per cent (ÖROK 2013, 40). In the coming years population growth, the transformation of family structures, and changing lifestyles will be among the main reasons for increased land consumption for residential purposes.

Apart from residential factors, other drivers of land consumption are business parks, office space, shopping, and the continuing expansion of recreational facilities. The development of commercial areas in Austria is a direct result of the local planning culture. Municipalities compete with each other, as they rely on tax income from businesses. From a financial point of view they must use all available options to entice businesses to settle within their perimeter, which is why Austria is dotted with an abundance of large and small commercial areas on the fringe of urban settlements.

Alongside commercial areas, retail developments in Austria have also achieved significant growth on the urban fringes. In total, the number of shopping centres rose from 160 to more than 200 between 2006 and 2012 (Standort und Markt 2012). The multiplication of shopping centres and specialist retail parks on the urban fringes has brought about strong spatial changes to cities and regions. Retail facilities that used to be found in city centres now nestle on their edges.

The fact that rural-urban fringes in Austria have developed into spaces where living, working, and shopping are found side by side means that many types of leisure facilities are available. Outdoor facilities include, above all, golf courses and sports grounds, which take up a lot of land. The number of golf courses increased dramatically between 2000 and 2012, from 108 to 152, which is a growth of 40 per cent (Statista 2013). In addition, there is a great range of other outdoor leisure and recreational areas, such as race-track driving, model airplane-flying fields, ski resorts with event areas for various festivals, and many others.

One characteristic shared by these facilities is good accessibility: many are situated near large cities or are located either near or right inside Austrian tourist destinations. In the western part of Austria, where tourism is more concentrated than in other regions, the density of such outdoor recreational areas is particularly high (Andexlinger et al. 2005).

A Case Study of the Vienna Region: A Boom of Out-of-Town Shopping Malls and Other Commercial Areas

Shortly before the outbreak of World War I Vienna had a population of 2.1 million. At the time it was assumed that the city would keep on growing, but owing to both world wars, expansion came to a halt in the first half of the twentieth century. Today, more than 1.8 million people live in Vienna and current development scenarios assume that the 2 million threshold will be exceeded by the year 2019 (Statistik Austria 2013). This growth is connected to the ongoing transformation of the city and its surrounding area. Evidence of urban dynamism can be seen in large development areas such as the former airfield in Aspern, where a new city district is being built and is set to become home to 20,000 people by 2028 (Stadt Wien 2014).

Until well into the 1990s the city's border areas had a residential character, as did the hinterland towns and villages from which many people commuted to Vienna for work, but this situation has changed. Shopping centres, specialist retail parks, trade fair centres, headquarters of top-ranking service companies, facilities for the health and education sectors, a few themed housing projects, and leisure facilities (golf courses, amusement parks, etc.) can now be found there (Figure 7.1). Owing

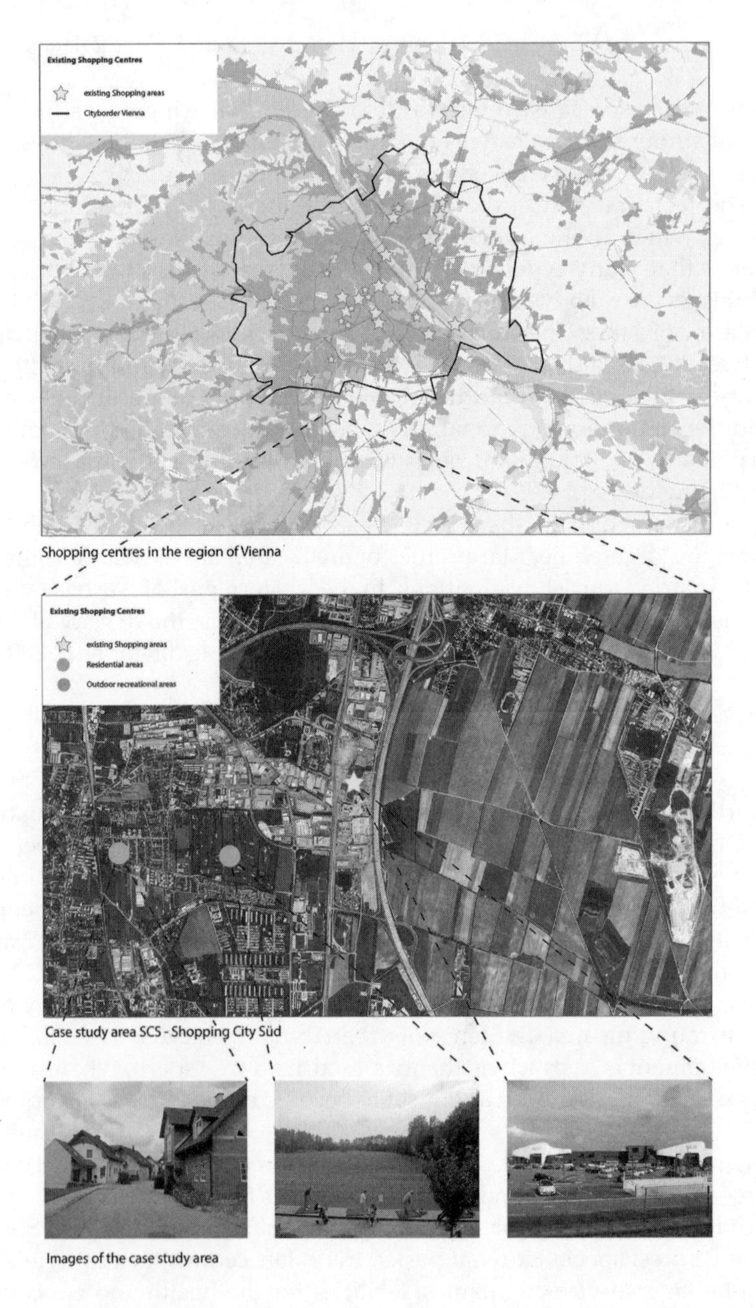

Figure 7.1 Vienna region, Austria. Maps and aerial pictures: EEA CORINE Land Cover, copyright not required. Photography and graphics: authors' copyright

to good transport connections to areas further from Vienna, these facilities are continuing to expand.

The City of Vienna has extensional links to the surrounding areas. Especially along the main transport corridors, large-scale developments of commercial areas and business parks can be found, as can many other building and use-typologies. The result is a stark, fragmented urban fringe.

Austria has some 201 shopping centres (Standort und Markt 2011). The largest one in terms of surface area is the Shopping City Süd (SCS) in the municipality of Vösendorf, to the south of the Vienna city boundary. With a total of 192,500 m^2 of rental space and 330 shops, it is also one of the largest centres in Europe. Every day some 50,000 cars are driven to the SCS (Seiss 2011). Vast car parks and large buildings have been constructed along the highway. Moreover, furniture stores, do-it-yourself stores, and a large exhibition space for prefabricated houses, hotels, and other facilities have established themselves in direct proximity to the SCS. Apart from shopping and business facilities, the SCS is surrounded by all sorts of recreational facilities (e.g., golf courses, lakes) and some residential areas. Altogether, it is an area characterized by diverse and self-enclosed functions.

A Case Study of North Tyrol: Leisure Facilities, Shopping Malls, and Residential Sprawl

North Tyrol is a region in the Alps that is characterized by a profusion of mountains and valleys. Its main valley (the Inntal) runs east-west and is about 150 km long. Its main population centre is the city of Innsbruck. There are also numerous small towns and villages, spreading to the east and west of Innsbruck or situated in the adjoining side valleys. The spatial development of settlements in North Tyrol has been shaped above all by small-scale political structures. There are 256 local authorities that – each for itself – steer the development of land use in the region.

Individual towns keep expanding in many parts of the Inn Valley, and it is no longer possible to say where one ends and the next one begins. Indeed, a settlement area has grown in this valley that very much resembles a linear or ribbon town; the whole North Tyrol region has even begun to be described in urban terms (Andexlinger et al. 2005). Over time the most diverse functions have spread themselves over the Inn Valley. A whole region has taken shape that, apart from its residential character, features out-of-town shopping malls and many other small commercial areas and leisure facilities, for instance, golf courses, event parks, sports grounds, and ski resorts, offering all kinds of additional recreational activities and equipment or theme parks (Figure 7.2).

Tyrol is a region whose topography is deeply marked by the Alps. Only about 13 per cent of the surface area is suitable for long-term settlement purposes. The Inn Valley, running along an east-west axis, constitutes the main settlement area, while side valleys feature only smaller settlements. The Innsbruck case study shows spatial development patterns that are dominated by residential interests and by commercial development, along transport corridors that are steered solely at the local level (within a municipality).

When one observes the parts of Innsbruck running eastwards – which also includes Thaur, Rum, and Hall – it becomes clear that settlement trends are driven by two factors: on the one hand, house building, specifically the construction of single-family houses; on the other, commercial sites alongside transport corridors. Both are contributing to a further expansion of the built-up area. On the whole, a fragmented settlement has emerged that is constantly growing and in which individual sites are merging into a structureless meshwork.

Planning Culture and Suburban Developments in the Netherlands

Spatial Planning in the Netherlands: Systemic and Integrated

The Netherlands is a country located in the delta of the rivers Rhine and Meuse. For many centuries the Dutch have been working to reclaim land from the sea and protect their countryside and cities from flooding. This goal could be achieved only through sustained collective efforts that made it possible to create a prosperous country in spite of such unfavourable geomorphological circumstances. These efforts form the basis for the way in which land and urban developments are organized in the Netherlands.

The Netherlands is about half the size (41,500 km^2) of Austria, whereas the Dutch population is almost twice as large (16.7 million inhabitants). This differential results in an average population density of about 400 inhabitants per km^2, which is four times higher than the average population density in Austria. Almost one-third of the inhabitants live in the western part of the Netherlands, called the Randstad, where the four largest cities (Amsterdam, Rotterdam, The Hague, and Utrecht)

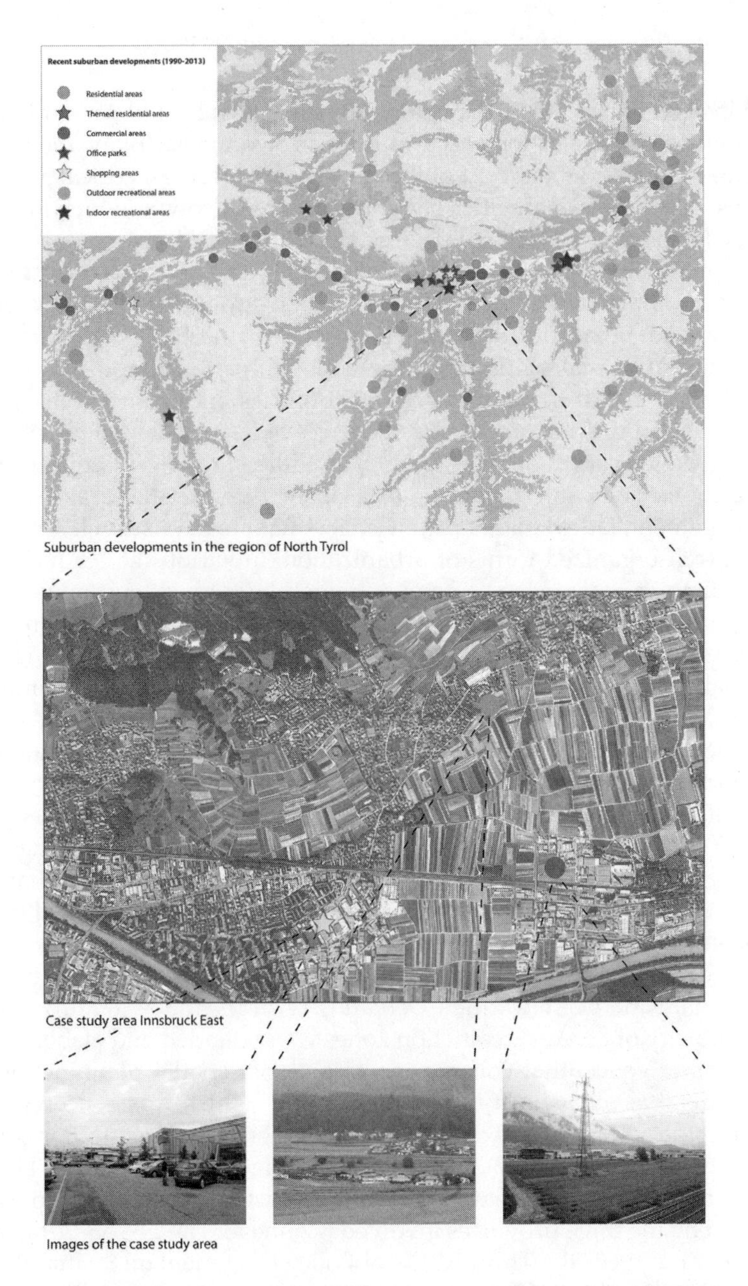

Figure 7.2 North Tyrol region, Inn Valley, Austria. Maps and aerial pictures: EEA CORINE Land Cover, copyright not required. Photography and graphics: Authors' copyright

are located. Since 1950 the population has grown by 170 per cent, and the built-up area has increased more than five times. Today, the built-up area covers 13.6 per cent of the country's surface. Historical growth maps of several Dutch cities show rapid urban growth, which has accelerated since the 1950s (Rutte and Abrahamse 2014).

As mentioned above, the specific natural conditions – a country located in a river delta – in combination with strong demographic, economic, and urban development helped form the basis for a distinct tradition of spatial planning at the national, regional, and municipal levels. In contrast to other European countries, including Austria, the national and regional authorities have played a very active role in the spatial development of cities since the middle of the twentieth century. In a context of limited space and faced with the challenges of water management, Dutch planners and policy-makers have striven for compact, well-organized forms of urbanization. In various Dutch national spatial policy documents different strategies designed to achieve compact urban forms have taken a prominent position (see Bartelds and De Roo 1995; Faludi and Van Der Valk 1994; Nabielek et al. 2012). These spatial policy documents have formed the basis for spatial planning and have influenced the way in which urban and suburban areas have developed in the Netherlands. In the following paragraphs the most important concepts and strategies will be briefly described.

In the years following World War II, planners and policy-makers already feared that cities would grow towards each other, leaving too little green space in between. Therefore, protected buffer zones that restrict urban development were introduced in the First National Policy Document on Spatial Planning (1960). The first two buffer zones were located between Amsterdam and Haarlem (Spaarnwoude) and between Rotterdam and Delft (Midden-Delfland). Over the following fifty years the strategy of "green" restriction zones was extended and elaborated.

Another conceptual cornerstone of national spatial planning since the early 1960s has been the "Green Heart," which was conceived to protect the rural area between Amsterdam, The Hague, Rotterdam, and Utrecht (Zonneveld 2007). Although the protection of the Green Heart was not supported by a national zoning regime, it gradually became accepted by the three provinces involved (Zonneveld and Evers 2014, 66).

After the 1960s the Third National Policy Document on Spatial Planning (VROM 1978) introduced the concepts of "clustered dispersal" and "growth centres," which were guided by the idea that expansion on the urban fringes could be limited by creating new towns that were located 10 to 30 km from the larger cities. Some of these settlements were completely new (e.g., Almere and Lelystad, to the northeast of Amsterdam), while others were linked to existing small towns or villages (e.g., Zoetermeer, to the east of The Hague).

At the beginning of the 1980s the concept of "clustered dispersal" came under growing criticism because new towns were dominated by residential features and showed a lack of urban qualities, such as density, cultural diversity, and mixed functions. Furthermore, the large cities – such as Amsterdam and Rotterdam – were faced with population decline and growing socio-economic problems. The reaction to this situation was to reorient policy on the existing large cities; this approach was framed by the concept of the "compact city." The focus shifted from urban expansion to revitalizing and densifying existing urban areas. The aim was to curb suburbanization and limit new urban development in the urban fringe.

Hence, the Fourth National Policy Document on Spatial Planning (VROM 1988) was based on the concept of the "compact city." However, this document did not manage to stop suburbanization. In the following period large-scale suburban neighbourhoods (called Vinex-locations) were planned as part of the supplement to the Fourth National Policy Document on Spatial Planning (VROM 1991). These new residential neighbourhoods were planned in a top-down manner and were designed to have relatively high densities and good access to public transport.

Nevertheless, the sheer scale of developments led to massive urbanization in some parts of the rural-urban fringe, especially in the Randstad. As a consequence, the population of inner-city areas continued to decline (K. Nabielek 2011). Moreover, strong economic growth encouraged municipalities to develop new commercial zones on the fringes of their cities. However, owing to a very restrictive policy concerning retail developments, the construction of large out-of-town shopping malls was prevented.

The most recent National Policy Document on Spatial Planning SVIR (I&M 2012) focuses on economic growth and large-scale infrastructural investment. Spatial planning has been decentralized to regional and local authorities, and national planning strategies, such as the national buffer zones, urban concentration, and densification, have been abolished. It can be expected that the liberalization and decentralization of spatial policy will accelerate urban development on the rural-urban fringe, and possibly beyond, depending on the policy decisions of regional and local planning authorities.

Recent Spatial Development in Suburban Areas in the Netherlands

Having considered conditions that are specific to the Netherlands in the previous section, in the next paragraphs we will describe suburban development since the 1990s. First, the built-up area has generally expanded between 1990 and 2010, from roughly 4,250 km^2 in 1990 to 5,750 km^2 in 2010, which equals an average growth of 75 km^2 per year, or 20 hectares per day. Accordingly, this growth is comparable to the daily growth of the built-up area in Austria (22.5 hectares per day). A study of urban development in the Netherlands between 1996 and 2003 has shown that the majority of growth in built-up areas has taken place in the urban fringe (Hamers et al. 2009): about 45 per cent of the new areas were residential, 40 per cent commercial, and about 15 per cent new recreational areas. In general, recent suburban residential areas are rather compact and located close to cities, whereas for new commercial and recreational areas the pattern is more fragmented (Nabielek, Kronberger-Nabielek, and Hamers 2013).

Compared with other countries in Europe, the scale and morphologic structure of the new suburban residential areas are remarkable (Boeijenga and Mensink 2008; Lörzing et al. 2006). The residential areas built in the recent past are primarily large-scale suburban neighbourhoods that were planned as part of the supplement to the Fourth Policy Document on Spatial Planning (VROM 1991). Some of the new neighbourhoods were planned for more than 30,000 inhabitants. Compared with suburban areas in other countries these Vinex-locations have a relatively high density (30 dwellings per hectare) and are predominantly made up of terraced houses. A concentration of large-scale residential areas can be found in and around the four largest cities in the Randstad: Amsterdam, Utrecht, The Hague, and Rotterdam. Examples of such residential neighbourhoods are Ypenburg (The Hague), Carnisselande (Rotterdam), Leidsche Rijn (Utrecht), and en Almere Buiten (Almere). Although there are a wide variety of housing typologies and styles, Vinex-locations have been criticized for their monofunctionality, repetitive layout, and inflexibility concerning future urban transformation tasks.

Next to these new suburban residential developments there has been strong growth of commercial areas such as business parks, office areas, industrial estates, and retail areas. Many of these commercial areas have been developed in the immediate vicinity of highway junctions and exits (Hamers and Nabielek 2006). They often fill up leftover strips between highway embankments and housing areas. The extent and character of this development has led to severe criticism. On the regional scale there have been complaints about the "filling up" of open spaces (mostly along infrastructure) and the diminishing contrast between city and countryside. On the local scale business estates are characterized by a very functional design and lack basic urban or architectural qualities. Furthermore, business estates without any connection to public transport have a negative impact on the environment, since they increase pollution and carbon emissions. Another problem is the growing number of decaying business estates with vacant lots and buildings (Buitelaar et al. 2013).

The growth of retail zones is relatively limited compared with that in other European countries because of a restrictive national policy concerning large out-of-town shopping areas, with the exception of large furniture stores, building supplies, and garden centres. The latter can often be found on the rural-urban fringe, especially in the vicinity of highways (Hamers and Nabielek 2006). Compared with the situation in Austria and many other European countries, where huge shopping malls can be found in urban peripheries (see the Vienna area discussion, above), this is a remarkable achievement of Dutch urban planning policy, and it has contributed to sustaining existing shops in central urban locations.

As mentioned above, about 15 per cent of the newly developed areas in the urban fringe are recreational. Recreational areas can have both a "red/urban" and a "green/rural" character. There is an essential difference between buildings with a recreational function (indoors) and outside areas with a recreational function (outdoors). Examples of the first group are ski halls, thermal baths, and mega cinemas; whereas examples of the second group include sports grounds, allotment gardens, natural areas, and parks. In the category of outdoor facilities, the growing number of golf courses is particularly striking. In the Netherlands the total area occupied by golf courses has increased from 1,300 to 7,300 hectares between 1998 and 2006 (Schuit et al. 2008). Many of the new golf courses are located in the vicinity of highways.

A Case Study of the Rotterdam-The Hague Region: Greenhouses, Business Parks, and Large-Scale Housing Areas

In the past twenty-five years the area between Rotterdam, Zoetermeer, and The Hague has been occupied by suburban developments consisting of recreational and commercial functions, greenhouses, and large-scale residential areas.

The Rotterdam-The Hague region can be described as a polycentric urban region containing two large cities. With 600,000 inhabitants, Rotterdam is slightly bigger than The Hague (about 500,000 inhabitants). Smaller cities in this region include Delft, Zoetermeer, Gouda, and Dordrecht. Furthermore, the region is characterized by a vast number of greenhouses used in the production of vegetables and flowers.

The map of urban developments in this region (Figure 7.3) shows that there has been almost no urbanization between the cities of Rotterdam and Delft. In this area the open landscape has been protected by the national buffer zone of Midden-Delfland that was established in 1960. The area between Rotterdam, Zoetermeer, and The Hague, on the contrary, has undergone fairly widespread urbanization – a combination of residential and commercial functions (mostly greenhouses). In these areas a new regional rail link has been established, connecting Rotterdam to The Hague. New residential neighbourhoods were constructed along this railway, which has led to a fragmented morphology: a scattered, widespread urbanization pattern that is quite unusual for the Netherlands. Furthermore, there have also been a number of new large-scale residential developments on the urban fringe in the Rotterdam-The Hague region. In these sites large, residential Vinex-locations were built on the "other" side of the highway. In these cases the highway constitutes a potent spatial barrier within the urban structure.

A Case Study of Flevoland: Themed Residential, Recreational, and Commercial Areas

Flevoland is one of the twelve Dutch provinces and is situated north-east of the city of Amsterdam. The area of Flevoland occupies former marshlands of the Zuiderzee that were disconnected from the North Sea by the construction of a major causeway in the 1930s. Subsequently, following a long tradition of reclaiming land from water bodies and marshlands, the Dutch began the construction of polders. Originally, they were used for agriculture. However, after World War II housing

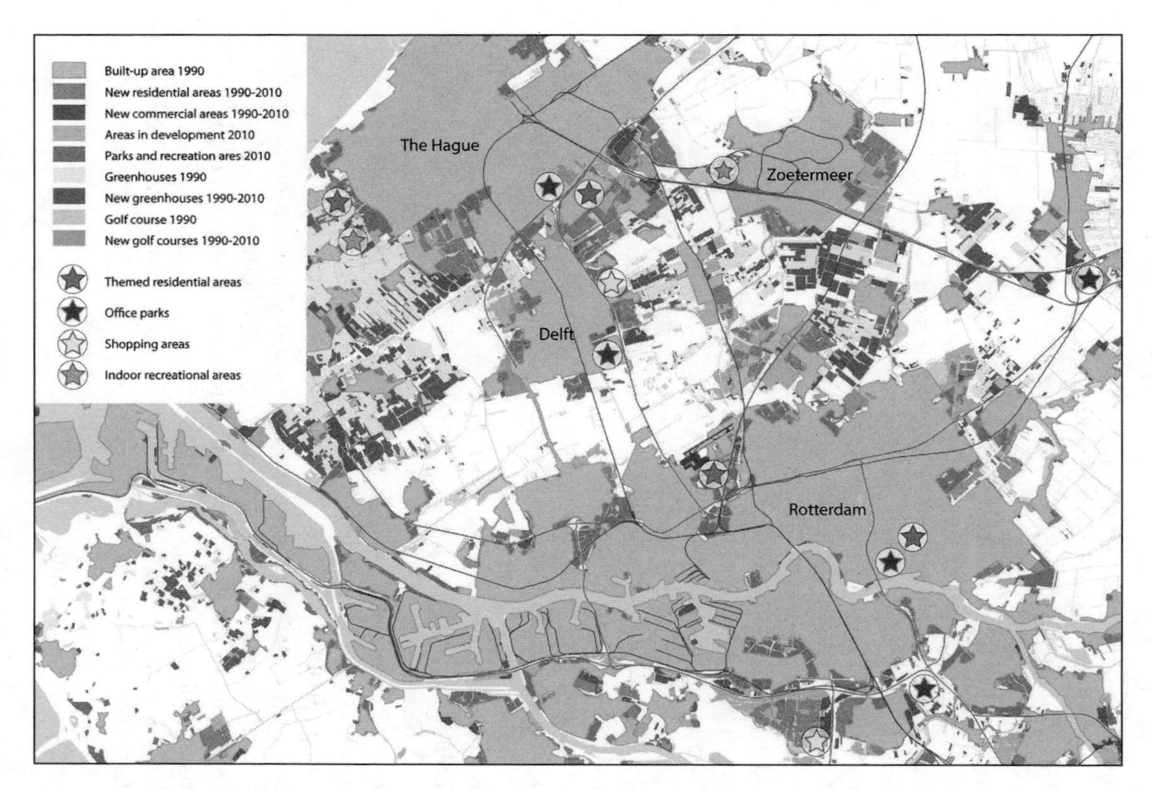

Figure 7.3 Recent suburban developments in the Rotterdam-The Hague region, Netherlands. Maps: CBS, edited by PBL, copyright permission obtained. Graphics: Authors' copyright

was needed for Amsterdam's rapidly growing population, and two new towns were planned in the polders: Lelystad and Almere (Figure 7.4). Their rapid growth (Almere is now the seventh-largest city in the Netherlands) put pressure on the national and regional road systems. A spatial mismatch between residential locations (e.g., in Almere) and employment locations (primarily in Amsterdam) increased the number of commuters and thereby the need for large-scale infrastructural investment, in both the railway and the highway systems.

The housing plan in Almere and Lelystad in the 1970s was centred on providing basic functionality and a levelling of social status. However, from the 1990s more exclusive homes began to be built with outstanding "identities." To attract new residents Almere and Lelystad became a laboratory for experimental housing typologies: environmental neighbourhoods, exclusive apartment "castles," residential areas that are combined with recreational functions (e.g. a golf course), and high-rise developments. Moreover, a large number of shopping and leisure facilities have been constructed there since the beginning of the 1990s.

The satellite picture of the western edge of Lelystad (Figure 7.4) shows a typical pattern of recent residential and recreational developments: new residential types, such as a large-scale recreational neighbourhood called Golf Park (a fusion of a golf course and single-family home area), a co-housing project that is inspired by Scandinavian community housing, and recreational homes with private boardwalks and boats in the harbour. Moreover, there are cultural and consumption spaces such as a museum and the large-scale factory outlet centre, Batavia Stad, in the style of a medieval village. The developments are set apart by their distinctive (themed) architectural styles and clear borders with their surroundings: fences, hedges, gates, and water bodies.

Almere and Lelystad are two new Dutch towns planned in the 1960s. The recent suburban fringe of Lelystad consists of a mixture of themed residential, recreational, and commercial areas that are spatially and functionally separated from each other.

Findings and Conclusions

In the past twenty-five years urban areas have expanded strongly in both Austria and the Netherlands. Moreover, the shape and composition of suburban areas have changed. Whereas suburban areas used to be dominated by residential and industrial functions, traditionally urban functions, such as shops, offices, and cinemas, have found their

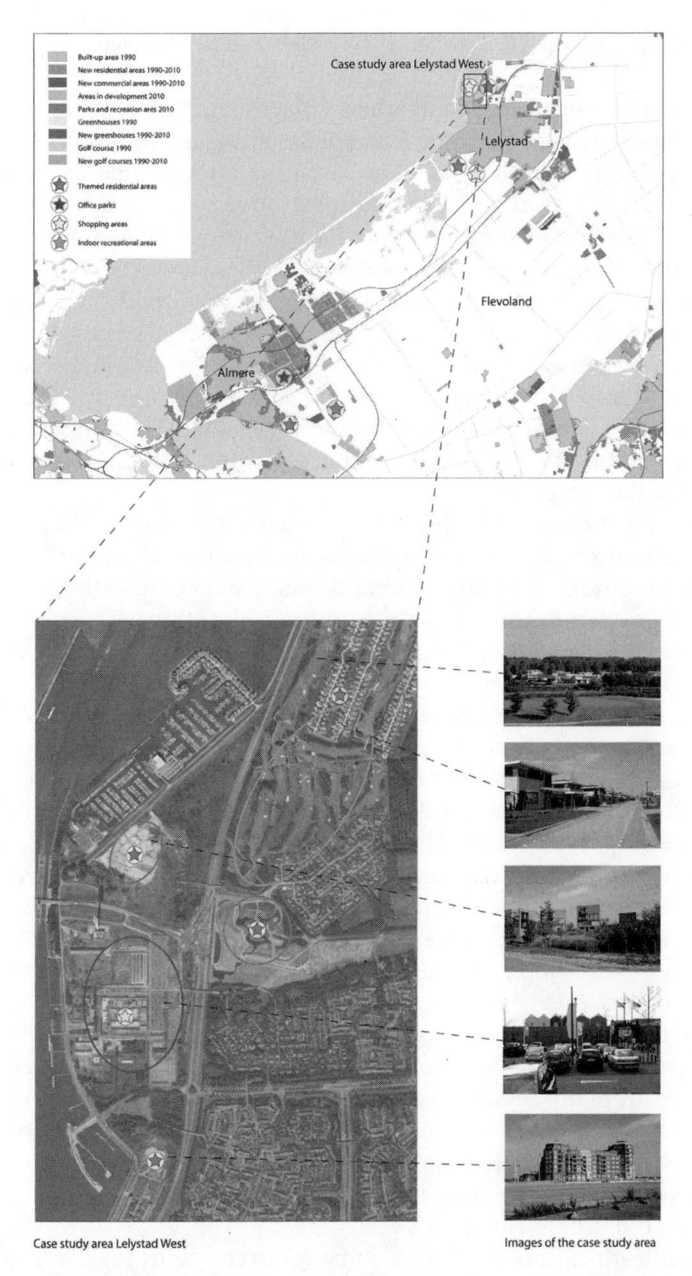

Figure 7.4 Recent suburban developments in Flevoland, Netherlands. Maps and aerial pictures: CBS, edited by PBL, copyright permission obtained. Photography and graphics: Authors' copyright

way into the periphery of cities and villages. Furthermore, recreational functions have become more widespread along the urban fringes.

However, the spatial analysis of suburban developments in Austria and the Netherlands and the analysis of four case-study areas lead us to conclude that beyond these similarities, there are some differences regarding the spatial characteristics of suburban areas. These can be ascribed to distinctive territorial and administrative conditions, but more specifically to very different planning cultures concerning the spatial domain. Whereas spatial development decisions are taken predominantly at a local level in Austria, the spatial planning system in the Netherlands can be described as comprehensive and integrated, having national frameworks that are delegated through the regional level down to the local level.

Looking at recent suburban developments in these two countries, we can see that the difference in planning cultures has certainly had an influence on the morphological and functional composition of the urban fringes. For example, a strong dissimilarity between new suburban residential areas is apparent. Whereas most new residential areas in Austria (with the exception of Vienna) have sprawled around existing urban areas in an organic way, new suburban residential areas in the Netherlands have been planned for specific locations in the vicinity of large cities. The latter tactic has led to large-scale suburban areas with relatively high population densities that enjoy good bicycle and public transport connections. The Dutch top-down approach has certainly proven to be very effective and has contributed to protecting green belts around cities. At the same time it has turned out to lack flexibility in the recent real estate crisis. Therefore, municipalities are currently looking for ways to develop residential areas in a way that is more organic and on a smaller scale.

The local approach in Austria seems to be more flexible but has led to fragmented urban structures and blurred borders between city and country. In many regions this more organic style has led to extensive land use, high car dependency, and visually unfavourable compositions. A more integrated planning approach with an effective regional framework would be necessary in order to protect rural and natural areas in the urban fringes.

Significant differences can also be observed with regard to out-of-town shopping and retail complexes. In Austria there has been strong growth of suburban shopping malls and retail centres. In combination with extensive parking services, these facilities consume large tracts of land, generate additional car traffic, and compete with shops in city centres. In the Netherlands, however, a very restrictive national retail policy has been effective in preventing the mushrooming of out-of-town shopping and retail areas, with the exception of some specific functions, such as large garden centres and furniture stores. Even though the national retail policy was abolished in 2003, provinces collectively decided to maintain the previous restrictive policy. Until now Dutch retail policy has successfully managed to retain shopping functions within central urban areas.

However, we can also observe certain similarities in suburban developments despite the different planning cultures in Austria and the Netherlands. In both countries business estates, office parks, and recreational areas – such as golf courses – have developed in similar ways. Business parks have emerged in the urban fringes, especially in the vicinity of highways and infrastructural nodes. In the Netherlands a national planning policy concerning business developments was unsuccessful because economic interests prevailed. Thus, decisions about business developments were taken mostly at the municipal level, which is comparable to the situation in Austria. As a result, suburban business developments in both countries are rather fragmented and disconnected.

To conclude, we can assert that planning policy certainly is one of the factors that has had an influence on the scale, shape, and functional composition of suburban developments. In correlation with other influencing factors (such as territorial, socio-economic, and institutional conditions) comparative studies of planning cultures and spatial developments in different countries can contribute to a better understanding of suburban developments. After all, a better understanding of the factors that affect urbanization is a basic prerequisite to being able to create more sustainable urban and suburban areas in the future.

REFERENCES

Andexlinger, W., P. Kronberger, K. Nabielek, S. Mayr, C. Ramiere, and C. Staubmann. 2005. *TirolCITY: New Urbanity in the Alps - Neue Urbanität in den Alpen*. A project by YEAN. Vienna: Folio Verlag.

Bartelds, H.J., and G. De Roo. 1995. *Dilemma's van de compacte stad: Uitdagingen voor het beleid*. The Hague: Vuga.

BEV – Bundesamt für Eich- und Vermessungswesen. 2009. *Regionalinformation der Grundstücksdatenbank*. 1 January 2009.

BEV – Bundesamt für Eich- und Vermessungswesen. 2011. *Regionalinformation der Grundstücksdatenbank*. 1 January 2011.

BMLFUW. 2002. *The Austrian Strategy for Sustainable Development*. Federal Ministry of Agriculture, Forestry, Environment and Water Management, Austria.

Boeijenga, J., and Mensink, J. 2008. *Vinex Atlas*. Rotterdam: 010 publishers.

Brenner, N. 2014. Introduction: "Urban Theory Without an Outside." In *Implosions/Explosions: Towards a Study of Planetary Urbanization*, ed. N. Brenner, 14–30. Berlin: Jovis.

Bryant, C., L. Russwurm, and A. McLellan. 1982. *The City's Countryside. Land and Its Management in the Rural-Urban Fringe*. London: Longman.

Buitelaar, E., N. Sorel, F. Verwest, F. Van Dongen, and A. Bregman. 2013. *Gebiedsontwikkeling en commerciële vastgoedmarkten: een institutionele analyse van het(over) aanbod van winkels en kantoren*. The Hague, Amsterdam: PBL Netherlands Environmental Assessment Agency / Amsterdam School of Real Estate.

Campi, M., F. Bucher, and M. Zardini. 2000. *Annähernd perfekte Peripherie: Glattalstadt / Greater Zürich Area*. German Edition. Basel: Birkhäuser.

Eisinger, A. 2005. "Einleitung: Stadtland Schweiz." In *Stadtland Schweiz: Untersuchungen und Fallstudien zur räumlichen Struktur und Entwicklung in der Schweiz*, ed. A. Eisinger and M. Schneider, 7–36. 2nd ed. Avenir Suisse, Basel, Zürich: Birkhäuser. https://doi.org/10.1007/3-7643-7660-0.

Evers, D. 2004. *Building for Consumption: An Institutional Analysis of Peripheral Shopping Centre Development in Northwest Europe*. Amsterdam: University of Amsterdam.

Faludi, A., and A. Van Der Valk. 1994. *Rule and Order: Dutch Planning Doctrine in the Twentieth Century*. Dordrecht: Kluwer Academic. https://doi.org/10.1007/978-94-017-2927-7.

Frijters, E., D. Hamers, R. Johann, J. Kürschner, H. Lörzing, K. Nabielek, R. Rutte, P. Van Veelen, and M. Van Der Wagt. 2004. *Tussenland*. Rotterdam and The Hague: nai010 publishers/RPB.

Gallent, N., J. Andersson, and M. Bianconi. 2006. *Planning on the Edge. The Context for Planning at the Rural-Urban Fringe*. London: Routledge.

Graham, S., and S. Marvin. 2001. *Splintering Urbanism: Networked Infrastructures, Technological Mobilities and the Urban Condition*. London: Routledge.

Hamers, D., and K. Nabielek. 2006. "Along the Fast Lane: Urbanisation of the Motorway in the Netherlands." *Proceedings of the International Forum on Urbanism 2006 in Beijing: Modernization and Regionalism: Re-inventing Urban Identity*. Vol. 1. Delft: IFoU, 274–81.

Hamers, D., and R. Rutte. 2008. "Shadowland. A New Approach to Land-in-between." (IHAAU) *Ezelsoren: Bulletin of the Institute of History of Art, Architecture and Urbanism* 1 (2): 61–80.

Hamers, D., K. Nabielek, M. Piek, and N. Sorel. 2009. *Verstedelijking in de stadsrandzone. Een verkenning van de ruimtelijke opgave*. The Hague: PBL Netherlands Environmental Assessment Agency.

I&M, Dutch Ministry for Infrastructure and Environment. 2012. *Structuurvisie Infrastructuur en Ruimte*. The Hague: I&M.

Lörzing, H., W. Klemm, M. Van Leeuwen, and S. Soekemin. 2006. *Vinex! Een morfologische verkenning*. Rotterdam and The Hague: nai010 publishers/RPB.

Nabielek, K. 2011. "Urban Densification in the Netherlands: National Spatial Policy and Empirical Research of Recent Developments." *Proceedings of Global Visions: Risks and Opportunities for the Urban Planet*, 5th Conference of the International Forum on Urbanism. Singapore: IFoU.

Nabielek, K., P. Kronberger-Nabielek, and D. Hamers. 2013. *The Rural-Urban Fringe in the Netherlands: Recent Developments and Future Challenges*. SPOOL. Accessed 19 September 2014. http://spool.tudelft.nl/index.php/spool/article/view/624.

Nabielek, K., S. Boschman, A. Harbers, M. Piek, and A. Vlonk. 2012. *Stedelijke verdichting: een ruimtelijke verkenning van binnenstedelijk wonen en werken*. The Hague: PBL Netherlands Environmental Assessment Agency.

Nabielek, P. 2011. "Theme Housing in the New Dutch Suburban Landscape." *Proceedings of Global Visions: Risks and Opportunities for the Urban Planet*, 5th Conference of the International Forum on Urbanism. Singapore: IFoU.

Rutte, R., and J.E. Abrahamse. 2014. *Atlas van de verstedelijking in Nederland: 1000 jaar ruimtelijke ontwikkeling*. Bussum: THOTH.

Schuit, J., H. Van Amsterdam, M. Breedijk, L. Brandes, E. Fick, and M. Spoon. 2008. *Ruimte in cijfers 2008*. The Hague: RPB.

Secchi, B. 1997. Un 'interpretazione delle fasi più recenti dello sviluppo urbano: la formazione della città diffusa e il ruolo dell infrastrutture. *Urbanistica Dossier* 3: 7–11.

Seiss, R. 2011. "Kommune und Raumplanung." In *Baukulturreport*, 76–85. Vienna: Bundeskanzleramt Österreich.

Sieverts, T. 1999. *Zwischenstadt: Zwischen Ort und Welt, Raum und Zeit, Stadt und Land*. Bauwelt-Fundamente Stadtplanung/Urbanisierung. Vol. 118, *3., verb. und um ein Nachw. erg. Aufl.* Braunschweig, Wiesbaden: Vieweg.

Stadt Wien. 2014. *Flugfeld Aspern*. Accessed 9 January 2014. https://www.wien.gv.at/wiki/index.php/Flugfeld_Aspern.

Standort und Markt. 2011. *Shopping Center Österreich*. Baden.

Standort und Markt. 2012. *Shopping Center Österreich*. Baden.

Statista. 2013. *Anzahl der Golfplätze in Österreich von 2000–2012*. Accessed 15 September 2013. https://de.statista.com/statistik/daten/studie/218257/umfrage/anzahl-der-golfplaetze-in-oesterreich/.

Statistik Austria. 2013. *Bevölkerung Österreichs seit 1869 nach Bundesländer*. Accessed 27 August 2013. http://www.statistik.at/web_de/statistiken/index.html.

Statistisches Handbuch Bundesland Tirol. 2014. Amt der Tiroler Landesregierung. Accessed 9 September 2014. https://www.tirol.gv.at/fileadmin/themen/statistik-budget/statistik/downloads/Statistisches_Handbuch_2014.pdf.

Tötzer, T., W. Loibl, and K. Steinnocher. 2009. *Flächennutzung in Österreich. Jüngere Vergangenheit und künftige Trends*. Wissenschaft und Umwelt Interdisziplinär. Nr. 12, 9–21. Vienna: Forum Wissenschaft und Umwelt. https://www.researchgate.net/publication/231492242_Flachennutzung_in_Osterreich_Jungere_Vergangenheit_und_kunftige_Trends.

Umweltbundesamt. 2013a. "Raumentwicklung." In *Zehnter Umweltkontrollbericht*. Vienna: Umweltbundesamt GmbH.

Viganò, P. 2001. *Territori della nuova modernità Provincia de Lecce, Assessorato alla gestione territotiale: Piano territorial do coordinamento = Territories of a New Modernity*. Naples: Electa.

VROM, Dutch Ministry for Housing and Environment. 1978. *Derde Nota over de Ruimtelijke Ordening*. The Hague: Sdu Uitgeverij.

VROM, Dutch Ministry for Housing and Environment. 1988. *Vierde nota over de ruimtelijke ordening*. The Hague: Staatsuitgeverij.

VROM, Dutch Ministry for Housing and Environment. 1991. *Vierde Nota over de ruimtelijke ordening extra, deel 1*. The Hague: Ministerie van VROM.

Young, D., P. Burke Wood, and R. Keil. 2011. *In-Between Infrastructure: Urban Connectivity in an Age of Vulnerability*. Accessed 1 July 2013. http://www.praxis-epress.org/availablebooks/inbetween.html.

Webber, M. 1998. "The Joys of Spread-City." *URBAN DESIGN International* 3 (4): 201–6. https://doi.org/10.1057/udi.1998.26.

Zonneveld, W. 2007. "A Sea of Houses: Preserving Open Space in an Urbanised Country." *Journal of Environmental Planning and Management* 50 (5): 657–75. https://doi.org/10.1080/09640560701475303.

Zonneveld, W., and D. Evers. 2014. "Dutch National Spatial Planning at the End of an Era." In *Spatial Planning Systems and Practices in Europe*, ed. M. Reimer, P. Panagiotis, and H. Blotevogel, 61–82. Routledge.

chapter eight

Factors Affecting Development Patterns in the Suburbs of Small to Mid-Sized Canadian Cities

JILL L. GRANT

Canadian planning practice in many cities reflects the growing influence of ideas originally associated with new urbanism and smart growth, but also described as linked with theories of sustainable development and healthy communities. Contemporary planning principles advocate denser, more compact, and more mixed suburban environments than are seen in the conventional developments of the post-war period. While principles promoting compact form, mixed use, and transportation alternatives appear widely in planning policies across Canada – in cities of all sizes and locations – the ability of planners to implement such ideas varies considerably. Consequently, the degree of spatial concentration (density versus sprawl) varies. Canada's largest cities – especially Toronto (Bunce 2004; Searle and Filion 2011) and Vancouver (Punter 2003; Quastel, Moos, and Lynch 2012) – have experienced increasing densities in the last few decades as they accommodated rapid growth resulting from immigration. Even cities such as Calgary and Edmonton – once renowned for low-density sprawl – have moved towards greater intensification and mix in new suburbs (Taylor, Burchfield, and Kramer 2014). The City of Ottawa has pursued an intensification strategy for over a decade (Leffers and Ballamingie 2013). However, not all cities have been as successful in transforming development outcomes (Filion 2010; Grant 2009).

A review of development experience in Canadian small to mid-sized cities illustrates a range of outcomes. Some smaller cities – especially those immediately adjacent to fast-growth areas – may be developing their suburbs at densities and in patterns comparable to those seen in core areas (Gordon and Vipond 2005). However, mid-sized cities in isolated, slow-growth regions generally retain a relatively low-density, segregated suburban pattern, even though the mix of uses may be changing (Leo and Brown 2000; Maoh, Koronios, and Kanaroglou 2010; Millward 2002). Smaller cities within commuting distance of fast-growth areas seem likely to straddle the extremes. A detailed investigation of trends in a range of mid-sized cities reflecting these differences can offer useful insights for those interested in how planning ideas get implemented. Here we discuss six smaller cities in four provinces. Except for Halifax, Nova Scotia, which is relatively isolated on the east coast, all are in fast-growth regions. Markham, Ontario, Surrey and Langley, British Columbia, border large cities: Toronto and Vancouver, respectively. Barrie, Ontario, and Airdrie, Alberta, are within commuting distance of Toronto and Calgary, respectively. We explore the relationship between growth rate and suburban pattern and three categories of effects: physical (relative location, residential geography, and size of city), economic (local economy, land market, and the geography of jobs), and social (cultural values, policy context, and demographic profile). We thereby hope to offer insights into some of the challenges planners face in applying new urbanism and smart-growth principles in suburban environments in second-tier cities.

Suburban Planning in Theory and Practice

For most of the period between World War II and 1990 the "garden city" model dominated city and suburban planning in Canada. Advanced by Ebenezer Howard (1985 [1902]), garden city planning called for blending the best of town and country in residential areas with appropriate amenities and services to allow working people to enjoy comfortable domestic lives in healthy environments in self-contained communities (Creese 1966). Over the decades, however, Canadian practice produced sprawling suburbs characterized by increasing levels of conformity, land-use segregation, and commuting by automobile (Harris 2004; Sewell 2009). By the 1970s planners in Toronto and Vancouver were urging intensification, containment, and mixed use (Grant 2006; Isin and Tomalty 1993; Punter 2003). Over the next few decades Canadian planning experienced a paradigm shift as philosophies associated with new urbanism and smart growth supplanted the garden city approach (Grant 2003, 2006). By the beginning of the twenty-first century cities large and small were preparing policies and plans advocating compact new development with mixed uses, mixed housing types, and walkable streets.

Closely associated with the work of Andres Duany and Peter Calthorpe, new urbanism employed an urban template for suburban development (Katz 1993). New urbanists recommended building higher densities, mixed uses, mixed housing types, a quality public realm, and walkable neighbourhoods. By the late 1990s proponents of smart growth had adopted new urbanism principles while emphasizing regional land-use planning and transportation alternatives (O'Neill 2000). By the late twentieth century Canadian planners saw the suburbs – once viewed as a solution to the woes of the industrial city – as problematic. New urbanism provided a theory for reshaping suburban growth. Following the lead of Toronto and Vancouver in trying to contain sprawl and enhance liveability, cities such as Calgary, Alberta, and Markham, Ontario, began exploring new urbanism in the 1990s. Calgary planners published the *Sustainable Suburbs Study* in 1995 and encouraged the development of McKenzie Towne, the first new urbanism community in Canada (Calgary 1995; Grant 2006). Markham worked with the Ontario provincial government to plan demonstration projects and to rewrite regulations to adopt new urbanism as its planning approach (Grant 2006; Lehrer and Milgrom 1996). Construction began on Cornell, the largest Canadian new urbanism community, in the mid-1990s. With enthusiastic support from the planning profession through conferences and publications, new urbanism principles promoting increased residential densities, mixed uses, and mixed housing types became entrenched as the common sense of the discipline by the end of the first decade of the twenty-first century. Eager to grow and enhance their urbanity, smaller cities consciously emulated policies they perceived as successful in larger ones. In the process of trying to achieve the objectives of new urbanism and smart growth in suburban areas, however, planners have often encountered challenges to implementation (Filion 2009, 2010; Filion and McSpurren 2007; Grant 2009).

Mid-sized cities experience particular challenges that larger cities may not encounter. Their economies are less diverse and robust than those of larger cities, making political leaders potentially less receptive to innovative policies and practices (Bunting et al. 2007). Cities located in high-growth areas face opportunities and challenges different from those in slow-growth regions (Bunting, Filion, and Priston 2002; Langlois 2010). In the next section I present recent research on six mid-sized and smaller Canadian cities in four provinces to explore the factors that influence the ways in which communities take up and try to implement planning principles associated with new urbanism and smart growth.

Practice in Six Cities

This chapter reports on a recent multi-year mixed-methods study focused on mid-sized and smaller communities near major cities: we examined cities in Alberta, British Columbia, and Ontario to understand how proximity to high-growth cities affected policies and outcomes in adjacent and commuting communities. The research involved field surveys, document review, census analysis, and interviews (with planners, councillors, developers, and residents) in each community. For comparison purposes we investigated a regional hub in Nova Scotia, to understand how relative isolation may influence suburban patterns and challenges.

The cities studied ranged in size from Airdrie, with just over 42,000 people, to Surrey, with over 460,000 people in 2011 (Table 8.1). Of the communities examined, Halifax grew most slowly: only 4.7 per cent from 2006 to 2011. By contrast, Airdrie increased by 46.2 per cent in the same five-year period. Densities varied from a low of 67.8 people per km^2 in Halifax – an amalgamated regional municipality the size of Prince Edward Island – to a high of 1245.3 per km^2 in Surrey, on the suburban fringe of Vancouver.

As Table 8.2 illustrates, the 2006 census indicated that the six cities accommodated different kinds of people. Markham and Surrey had high proportions of immigrants, largely from Asia. By contrast, Airdrie and Halifax had rates of immigration significantly below the national average. The communities varied widely in household size: the average household in Halifax had one fewer person than did an average household in Markham. The median age in Langley District was 6.5 years older than the average in Airdrie. Incomes ranged from just above the national average in Halifax to almost 50 per cent above the national average in Markham. In April 2014 regional unemployment rates ranged from a low of 5.3 per cent in Calgary, near Airdrie, to a high of 7.8 per cent in Toronto, next door to Markham; only the Ontario communities had unemployment rates above the national average (Canadian Press 2014).

Housing markets differ considerably across the communities, although all have experienced inflation in property values over the last decade. Table 8.3 indicates that in 2010 home values in Markham, Surrey, and Langley were more than twice the median price of a detached home in Halifax. Housing markets in all the cities except Halifax

Table 8.1. Population growth and density, Canadian cities

Place	Population, 2011	Growth (%) 2006–11	Density per km² 2006
Halifax	390,328	4.7	67.8
Markham	300,135	14.7	1230.5
Barrie	187,013	5.6	197.3
Airdrie	42,280	46.2	874.0
Surrey	463,340	17.3	1245.3
Langley	103,145	10.0	305.4
Canada	33,476,688	5.9	3.5

Notes: Halifax refers to the Halifax Census Metropolitan Area. Barrie refers to the Barrie Census Metropolitan Area. Langley refers to the Langley District Municipality.

Sources: Statistics Canada (2008, 2013, 2014)

Table 8.2. Population characteristics, 2006, Canadian cities

Place	Immigrants % of total, 2006	Average household size, 2006	Median age, 2006	Median household income in CAD, 2006
Halifax	7.4	2.4	39.0	54,129
Markham	56.5	3.4	38.1	79,924
Barrie	12.8	2.7	36.7	64,832
Airdrie	6.8	2.9	32.6	78,097
Surrey	38.3	3.0	37.0	60,168
Langley	17.1	2.8	39.1	69,805
Canada	19.8	2.5	37.2	53,634

Source: Statistics Canada (2008)

are in some ways influenced by the development context of large, dynamic neighbouring cities, which drive up demand, exacerbate transportation costs, and increase home prices. The proportion of detached housing in each city offers a useful indicator of housing mix. Detached houses made up only 43.8 per cent of the market in Surrey, where multi-family units are common. By contrast, Airdrie is a community of detached homes: 72.6 per cent of the housing stock. The next sections briefly describe each community in turn.

Table 8.3. Housing characteristics, Canadian cities

Place	Median price single-detached house in CAD, 2010	Detached housing as % of total, 2006
Halifax	251,116	51.6
Markham	539,990	67.3
Barrie	368,695	70.4
Airdrie	389,700	72.6
Surrey	610,000	43.8
Langley	699,000	61.2
Canada	344,782	55.3

Sources: CMHC (2011, 2013); Statistics Canada (2008, 2014)

Halifax, Nova Scotia: A Regional Hub in a Slow-Growth Region

Home to around 390,000 people in 2011, Halifax is the provincial capital of Nova Scotia, and the Maritimes economic hub (Brender and Lefebvre 2010). The economy depends primarily on the government, military, healthcare, knowledge services, and logistics associated with port activity. A $25 billion ship-building project contracted with the national government in 2013 now offers the promise of an economic boom after years of growth below the national average. Local leaders hope that growth will ameliorate the effects of low wages, limited immigration, an aging population, and small household sizes.

In 1996 the province amalgamated Halifax with other small cities and Halifax County to create a regional municipality with a huge area: almost 5500 km^2. Development is concentrated around Halifax harbour, but suburban growth is scattered at relatively low densities in several directions from the core along roads, lakes, and ocean coastline (Millward 2002). Although the regional plan (HRM 2006) promoted intensification and mixed use, plans and approvals that predated the regional plan's adoption stymie efforts to regulate sprawling suburban development (Brewer 2014). With abundant opportunity to find relatively affordable housing in urban, suburban, or rural settings within easy commuting distance of employment centres, Halifax has the lowest density and average house prices of the six cities studied. New suburban districts have a wider mix of uses and housing types than older ones do, but development patterns remain highly segregated by land use and housing type (Perrin and Grant 2014) and consistently low density (Brewer and Grant 2015). Defying expectations, however, Halifax has a relatively low proportion of single-detached housing. Although high land values and commuting distances push up demand for multi-family housing in places such as Surrey, low average incomes and large numbers of university students and military personnel may explain the robust market for multi-family and row housing, even in suburban Halifax (Figure 8.1).

Markham, Ontario: New Urbanism Exemplar

Home to almost 300,000 people in 2011, the Town of Markham has grown rapidly over the past few decades from a small town to a substantial suburban municipality within commuting distance of Toronto. At 1230.5 people per km^2, Markham is denser than most of the communities examined. Gordon and Vipond (2005) found higher densities in newer suburbs built to new urbanism principles, although Skaburskis (2006) argued that the gains were modest and may not be sustained. The area is within the region designated for intensification by Ontario's "Places to Grow" legislation. With a large concentration of high-tech companies and highly educated workers, Markham is the most affluent of the communities examined: the median household income in Markham was almost 50 per cent higher than the Canadian average in 2006. Average household size in Markham is substantially larger than in the other communities: multi-generational households among the large immigrant population (56.5 per cent) likely explain the difference. Some 46 per cent of households include couples with children, almost 60 per cent higher than the Canadian average (Statistics Canada 2008). Markham is arguably the most diverse community in Canada, with over 65 per cent of residents having identified themselves as visible minorities in the 2006 census: as Grant and Perrott (2009) noted, however, new residents have self-selected to neighbourhoods in ways that reproduce ethnic and class segregation.

Markham is certainly the most successful of the six communities in achieving new urbanism principles. By integrating the principles into its planning policies and regulations it has created the infrastructure to require higher densities, land-use mix, and mixed-housing types (Perrott 2008). Its suburban neighbourhoods have a distinctly urban character, with small but attractive homes on grid streets (Figure 8.2).

Figure 8.1 Multi-family units in a Halifax, Canada, suburb.
Photo: Kirk Brewer, used with permission

Figure 8.2 Attractive row housing in Markham, Canada.
Photo: Katherine Perrott, used with permission

High demand for housing in a context with a diverse consumer market has allowed Markham to carve a niche for dense residential neighbourhoods of traditional-style homes on parks and relatively narrow but walkable streets. Ethnic clustering around community services (such as mosques or churches) or amenities (such as shops) has reinforced the market for high-density (sub)urbanism.

Barrie, Ontario: Small Town or Distant Suburb?

Located about one hour's drive north of Toronto on Lake Simcoe, the Barrie CMA had about 187,000 people in 2011. Its growth moderated in the 2006 to 2011 period (5.6 per cent), after phenomenal (19.2 per cent) expansion from 2001 to 2006 (Statistics Canada 2008). Its population is younger than the Canadian average, and median household income is $11,000 above the national average. The average house price in Barrie is about two-thirds of what the average home costs in Markham, an hour south: residents we interviewed often noted that they could get more land and house for their money in Barrie than they could in suburbs near Toronto. Barrie's southern suburbs have grown as a popular choice for those looking for larger homes and lots while working in the northern parts of the GTA: about one-third of the population commutes to work elsewhere. Go Transit service to Toronto is available, although not frequent: most residents rely on automobiles for transportation.

Its population distributed within the town and in smaller settlements around the district, Barrie's density is relatively low and its lots are large (Rosen and Brewer 2013). The provincial growth strategy (Ontario 2012) designated Barrie as a growth and intensification area. Although Barrie's 2009 Official Plan promoted intensification and growth within a sustainable development philosophy, concerns about the quality of the urban realm affirmed the influence of new urbanism in the plan. Amendments made in 2010 to strengthen the plan's adherence to the province's Places to Grow legislation reinforced smart-growth premises and indicated a commitment to increasing multi-family housing (Barrie 2010). Not surprisingly, Barrie is torn between its history as a small lakeside town, its present as a far-flung commuter suburb of Toronto, and its future wish to be a dense and sustainable city. At present, many of its suburbs feature conventional patterns with large detached homes (Figure 8.3).

Figure 8.3 New suburbs in Barrie, Canada, have a conventional feel. Photo: Gillad Rosen, used with permission

Airdrie, Alberta: The "Canadian Dream" outside Calgary

Located 35 km north of Calgary, Airdrie has annexed land to keep up with vigorous growth in the oil and gas sector in the last decade. A labour market survey in 2013 suggested that a majority of Airdrie workers commuted outside the community to work (Airdrie 2013). Its population grew by 46.2 per cent between 2006 and 2011 to just over 42,000. Although it is independent, it effectively functions as a commuter suburb for Calgary. Its density, at 874 people per km^2, is below that seen in the immediate peripheries of Vancouver and Toronto, but well above densities in the other three cities sampled. Airdrie has the youngest median age of the six communities and an average household larger than the Canadian norm. Some 43 per cent of its households are couples with children at home, well above the Canadian average of 29 per cent (Statistics Canada 2008). Its residents are affluent, with a median household income 46 per cent above the national average and 14 per cent over the Calgary median.

As part of the Calgary Regional Partnership, Airdrie seeks to align its plans to those of the Calgary Regional Plan. Its planning policy and

Figure 8.4 Higher-density housing options are becoming more common in Airdrie, Canada. Photo: Troy Gonzalez, used with permission

political leaders endorse new urbanism and smart-growth principles, calling for a mix of housing types and land uses (Airdrie 2009). However, residents appreciate conventional development forms. Many of those interviewed in Airdrie talked about the persistence of the "Canadian Dream" in Airdrie: living in a detached house with an attached garage. Developers active in the Calgary housing market also build homes in Airdrie. The kind of new urbanism projects seen at McKenzie Towne in Calgary do not appear in Airdrie, but development densities are rising (Figure 8.4) and some new housing forms are appearing (Gonzalez 2010). In 2012 Airdrie adopted a new sustainability plan, Airdrie One, to help coordinate its planning approaches (Airdrie 2012).

Surrey, British Columbia: The Rise of a High-Density Suburb

Until the 1950s Surrey was a farming and fishing community on the Fraser River outside Vancouver. Today Surrey connects to Vancouver by Skytrain and road. Growth accelerated during the 1980s and 1990s: between 2006 and 2011 the population grew by 17.3 per cent. With suburban growth areas constrained by the province's Agricultural Land Reserve policy, adopted in the 1970s (Hanna 1997), Surrey has the highest density per km^2 of the six communities. The median age is close to the national average, and the median household income is modest by suburban standards around Vancouver.

Surrey has a large immigrant population (38.3 per cent); about 30 per cent have south Asian roots (Statistics Canada 2008). Some suburbs include large Sikh communities, with temples and specialty retailers. Individual suburbs in Surrey have distinct histories and character (Sinoski 2013). Planning policy seeks to make the suburbs more urbanized by supporting town centres featuring a mix of uses and higher densities. Planners are proud of the East Clayton neighbourhood (Figure 8.5), a suburb designed with new urbanism and sustainability principles and featuring smaller homes on narrow streets. Surrey has the smallest proportion of detached housing of the cities studied, and the second-highest median house price. High land values in the Vancouver area have rapidly shifted the market to greater acceptance of multi-family housing options and are affecting home ownership rates.

Langley District Municipality, British Columbia: From Rural to Urban

On the eastern border of Surrey, on the commuting fringe of Vancouver, Langley District Municipality includes vast areas of farmland sprinkled with village and suburban clusters. Langley grew by 10 per cent between 2006 and 2011. Some 75 per cent of the area is designated Agricultural Land Reserve, thus concentrating urban development in dense nodes (Langley 2014) and in the City of Langley (which is excluded from figures reported for the township). Overall population density is towards the middle for the sampled communities, but for only one-third of Surrey's. The median age in Langley is highest in the sample, reflecting the popularity of the area as a retirement destination. Median household income is $11,000 above the Canadian average, but still $10,000 below Markham's.

Langley coordinates its planning activities with Metro Vancouver. The municipality affirms its commitment to new urbanism principles: "We pride ourselves on developing vibrant, complete neighbourhoods which offer flexible, affordable, and mixed housing options and the opportunity to live, work, shop, and play in a safe community" (ibid.). It has organized development in high-density town centres (often former villages or redeveloped suburbanized areas) where it can provide services and amenities. Although older forms of residential development in Langley featured detached housing on 2-acre lots, new communities

Figure 8.5 Developments such as East Clayton in Surrey, Canada, show New Urbanism principles in action. Photo: Dan Scott, used with permission

have multi-family units with relatively small building footprints and densities that may reach 80 units per acre (Scott 2010). Langley also has a significant concentration of private and gated communities (Figure 8.6) developed in the 1980s and 1990s (Grant and Curran 2007). The mix of housing forms creates stark contrasts.

Trends in Development in Smaller and Mid-Sized Cities

Based on examining communities in four provinces, what can we reasonably conclude about common development trends in the suburbs of smaller Canadian cities? First, policies and plans in all six communities share the commitment to increasing densities to enhance efficiencies, mixing land uses to make communities more self-contained, mixing housing types to accommodate a range of residents, and developing in ways that reduce automobile use. New urbanism and smart-growth principles informed policy in Markham by the mid-1990s, and by the mid-2000s they permeated planning documents in the other cities. Second, we see that practice is changing: more multi-family units are being built in the suburbs than during the 1970s and 1980s, condominiums or strata developments have come to the suburbs in considerable numbers, and suburban commercial centres address daily and

Figure 8.6 Langley's (Canada) higher-density suburbs often have private roads, sometimes with gates at the entry. Photo: Dan Scott, used with permission

occasional needs. In contemporary Canada the suburbs of smaller cities are sites of investment and growth and wield considerable political power. They are arguably more urban than the conventional suburbia of the last generation. In all the communities, however, weaning residents away from automobile transportation has proven a significant challenge: over three-quarters of commuters drove to work, the lowest proportion (76.6 per cent) being in Halifax and the highest in Langley (91.4 per cent). Even in Markham and Surrey, with their new urbanism developments, over 81 per cent of commuters drove to work in 2011 (Statistics Canada 2008).

Although we found similarities across the communities, we also identified differences. The circumstances of these cities vary widely, affecting their abilities to implement policies on intensification and land-use mix in new suburban developments. The greatest successes appeared in Markham, Surrey, and Langley: in these locations near Toronto and Vancouver provincial policy, rapid growth, and land scarcity coalesced to create economic and political conditions that have supported intensification and mixing uses and housing types. Smaller cities – such as Airdrie and Barrie – within longer commuting distances from major metropolitan areas, have had mixed results. Cultural preferences for detached houses remain strong in places like Barrie and Airdrie. While such cities have seen some success in developing new suburbs at higher densities, they have also annexed rural land to expand and continue to build large numbers of single-detached homes for automobile commuters. We discovered significant challenges to the new planning agenda in Halifax, a remote regional hub: despite efforts to initiate regional planning, suburban development there has struggled to overcome a legacy that enabled sprawl, restricted intensification, and reduced the viability of public transportation options. We also found that access to public transit varied widely across and within suburban communities.

What critical factors explain development patterns in Canadian urban environments? Although we find it analytically convenient to treat smaller or mid-sized cities as a set, significant differences persist among the places examined. Community size is not necessarily aligned with the rate or character of suburban development. Based on analysis of these six cities, we suggest that three categories of salient factors influence the type of suburban development that occurs. Physical, economic, and social factors interact to create the context within which suburban development occurs. We briefly discuss these factors in turn.

Physical Factors

Primary among the physical factors is relative location. Where a smaller city is located affects the character of suburban development and the pace of change. As Lepawsky, Hall, and Donald (2014) argue, relative location within the system of Canadian cities matters: a city like St John's, Newfoundland, isolated on an island far from other urban centres, experiences different pressures and opportunities from those of Kingston, Ontario, located between Toronto, Montreal, and Ottawa. Proximity to large cities may generate employment prospects, regional growth rates, and land costs not found in more remote cities. Thus, Markham experiences spillover from Toronto, while Halifax has only its own circumstances to drive growth and influence expectations. Relative location near booming resource industries – like the oil sands – will similarly influence the character and rate of development. Transportation connections, whether through roads or public transit systems, shape how and where cities may grow.

Space is not featureless. In each community the nature of local attractions, landscape features, and hazards affects the residential geography of where and how development occurs. Long-distance commuters may want to live in suburban areas closer to the transportation networks or systems they access to get to work. Particular amenities, such as coastlines, add value to some locations and thus affect the kinds of developments built there. Rugged areas, hills, lakes, and other geographic features influence the continuity of the urban fabric: where development is discontinuous, as in Halifax, achieving higher densities and land-use efficiencies proves extremely challenging. By contrast, tightly bounded communities with reasonably uniform geographies, such as Markham, may contain and link development more readily.

Although the size of a community is not the primary factor influencing development patterns, size does affect growth rates. If smaller cities capture suburban growth within high-growth regions, then their growth rates may accelerate rapidly. Thus, a small community like Airdrie expands at a phenomenal rate, even though it absorbs a relatively small proportion of Calgary-area growth.

Economic Factors

Local economic conditions certainly reflect the influence of relative location and growth rates. The nature of the regional economy affects the kinds of jobs available and consequently the level of affluence in the population. More affluent residents can make different kinds of choices than residents with lower incomes. Affluent areas like Markham and Langley enjoy stronger investment in the urban realm and in local amenities: their high-density environments appear more urban in character than new suburban nodes in lower-income areas such as Halifax. Lower average incomes in Halifax may explain the prevalence of rental housing in a community with abundant opportunities for suburban and exurban development. Provincial and local economic conditions also affect investment in alternative transportation options, such as transit systems and bikeways.

The nature of the local land market has a major impact on development outcomes. Where a few companies monopolize the local land market, pressure for intensification may increase, whereas greater competition may result in diversity of development options so that lower-density options are less likely to diminish. The growth of large development corporations in many markets, often producing housing on speculation, arguably increases homogeneity of suburban form. In any market the availability of mortgage credit or provincial housing programs may influence the rate at which buyers enter and the price

Plate 1a Single-family home development outside Mexico City, Mexico. Photo: Ute Lehrer, 2014

Plate 1b New subdivisions in geometric patterns on the outskirts of Calgary, Canada. Photo: Ute Lehrer, 2014

Plate 2a FlexSpace in the outskirts of Vienna, Austria. A typical expression of today's urbanization processes in the vicinity of Vienna, where freeway and railway lines dissect the urban landscape and where we find single-family homes and apartment buildings next to warehouses and industrial complexes, and where the remaining space is left for agricultural activities. Photo: Ute Lehrer, 2015

Plate 2b Infrastructure and housing is competing with agricultural land and industrial uses north of Beijing's urban area, China. Photo: Ute Lehrer, 2015

Plate 3a Warehouse and industrial activities in the Munich area, Germany. Photo: Ute Lehrer, 2008

Plate 3b Industries, warehouses, and housing on the outskirts of São Paulo, Brazil. Photo: Ute Lehrer, 2016

Plate 4a Construction activities financed by private investment in Greece. Photo: Roger Keil, 2015

Plate 4b Massive housing towers built by development companies in the Lingang area, Shanghai, China. Photo: Ute Lehrer, 2015

Plate 5a Informal settlements alongside a garbage dump on the east side of Mexico City, Mexico. Photo: Ute Lehrer, 2014

Plate 5b Informal settlements in Gauteng, South Africa. Photo: Ute Lehrer, 2015

Plate 6a Individual housing in the outskirts of Tokyo, Japan. Photo: Ute Lehrer, 2008

Plate 6b Individual housing in the outskirts of Athens, Greece. Photo: Ute Lehrer, 2015

Plate 7a Building into the forested area in Helsinki, Finland. Photo: Roger Keil, 2016

Plate 7b Defining an urban edge along the foothills in Mexico City, Mexico. Photo: Ute Lehrer, 2014

Plate 8a Massive and repetitive high-rise housing on the outskirts of Beijing, China. Photo: Ute Lehrer, 2015.

Plate 8b Geometric formation of mass housing in the Paris area, France. Photo: Ute Lehrer

point at which they can purchase homes. At the community level the relative cost of housing varies across developments: the choices purchasers make reflect their evaluation of the opportunities available in the market.

Concentrations of employment opportunities and transportation access to them influence the character and pace of development. Locations with good access to employment centres offer an advantage to prospective buyers. Sacrificing space or accommodating density increases may be perceived as a viable trade-off for shorter commute times or greater transportation options. Communities like Markham benefit from concentrated employment in high-tech sectors. Airdrie and Barrie are attractive to workers in industries connected to employment concentrations around the Calgary and Toronto airports, respectively.

Social Factors

Although Canadians may share cultural values at a high level, local differences in political behaviour, community characteristics, and residential preferences appear in the places evaluated. Local political cultures differ widely. For instance, political leaders in Markham generally supported new urbanism ideas from the 1990s on, and they transformed local policy to apply such ideas in new development. By contrast, fractious local politics in Halifax frustrated efforts to concentrate development in specific nodes. Some smaller cities have large ethnic communities that bring particular values to their expectations about urban form and function. In some of the cities – such as Airdrie, Barrie, and Halifax – the desire for large houses on large lots continues to provide strong markets for conventional suburbs. Consumer preferences to live with others like them and buyers' interest in natural features – such as waterfront living – continue to drive class segregation in suburban development. At the same time, however, growing acceptance of condominium or strata ownership has validated higher-density housing options in suburban markets. The willingness of local governments to permit condominium development on private streets has provided consumers concerned about privacy and exclusivity with mechanisms to regulate access to make higher-density living more culturally acceptable.

The regional and local policy context influences development outcomes. Plans and policies in all the communities encourage new urbanism principles, but the effectiveness of implementation mechanisms varies across the communities studied. Where political leaders legislate in ways that encourage intensification and mix, development patterns more quickly shift to reflect new urbanism principles. Thus, we see British Columbia and Ontario communities moving to higher densities and mix in new developments. Provincial regulations – the agricultural land reserve in British Columbia and the Places to Grow legislation in Ontario – create powerful tools to transform practice. Alongside high rates of growth and shifting consumer expectations, policy can create an environment within which higher densities and diverse housing types become integrated in the suburban market. Without appropriate tools and the right economic and cultural conditions, however, policy cannot shift practice as far as planners hope to see.

The demographic profile of the community can shape outcomes. The rate of growth in mid-sized cities depends on migration. Smaller cities near larger cities enjoy the spin-offs of immigration. The choices that some new immigrant communities have made to congregate in particular cities, such as Markham and Surrey, shape the character of development in those places. Higher densities and community-specific built forms (such as mosques) are common in some suburban locations where ethnic communities cluster. Communities that prove attractive to young families, such as Airdrie, may develop different suburban amenities than those, such as Langley, that recruit retired couples. In cities such as Halifax, where large numbers of rental units have been constructed in some suburban areas, less affluent households and single-parent families may be shifting residential preferences. Thus, the character of the population may influence development outcomes, including whether new urbanism design principles prove popular.

Areas that are growing rapidly experience greater development pressures and price increases that encourage innovation and intensification. In contexts of rapid growth, land developers may feel confident in experimenting with denser options to optimize returns on land. Tight housing markets encourage residents to consider more affordable (e.g., smaller) housing types and may make political leaders more receptive to regulatory changes.

Challenges to Change

While Canada's largest cities are seeing increasing suburban densities and land-use mix in current land markets, smaller cities sometimes struggle to achieve their aims. Structural barriers to change make transforming conventional development patterns challenging even in

Canada's largest cities (Filion 2010; Grant 2009). In mid-sized cities retail uses may fail to retain their viability in new mixed-use development areas (Grant and Perrott 2011). Finding ways to get people out of their cars can prove daunting even in higher-density suburbs. Smaller cities have followed the lead of Canada's large urban centres in putting policy in place to transform sprawling and segregated land uses into new patterns that result in compact form and complete communities. However, experience shows that achieving new development patterns in Canadian suburbs in smaller cities will happen quickly only where market conditions are appropriate and where higher levels of government create regional policies that contain sprawl and compel mix. Smaller cities near areas of rapid growth are thus proving more successful in transforming their development practices to conform to new urbanism and smart-growth philosophies than are mid-sized cities in more peripheral locations.

ACKNOWLEDGMENTS

Funding for this research came from grants from the Social Sciences and Humanities Research Council of Canada: Standard Research Grants (to Jill Grant) 410–2009–4 for "Trends in the suburbs" and 410–2006–3 for "Theory and practice in planning the suburbs"; Major Collaborative Research Initiative grant (to Roger Keil) for "Global suburbanisms: governance, land, and infrastructure in the 21st century (2010–2017)." I'm grateful to Katherine Perrott, Blake Laven, Gillad Rosen, Dan Scott, Troy Gonzalez, Heidi Craswell, Leah Perrin, and Kirk Brewer for research assistance between 2007 and 2014. Thanks to Richard Harris and Ute Lehrer for helpful suggestions on an earlier draft.

REFERENCES

Airdrie, City of. 2009. *Municipal Development Plan*. Accessed 3 August 2014. http://www.google.ca/url?sa=t&rct=j&q=&esrc=s&source=web&cd=1&ved=0CB4QFjAA&url=http%3A%2F%2Fwww.airdrie.ca%2FgetDocument.cfm%3FID% 3D1130&ei=oNnkU7LUNsyQyATGrYGwDw&usg=AFQjCNFmB2vUqMWToJ4AvTfQsWZUbM8M9g&bvm=bv.72676100,d.aWw.

Airdrie, City of. 2012. *Airdrie One Sustainability Plan*. Accessed 3 August 2014. https://www.airdrie.ca/index.cfm?serviceID=25

Airdrie, City of. 2013. *Labour Market*. Airdrie Now / Economic Development. Accessed 3 August 2014. https://www.airdrie.ca/index.cfm?serviceID=481

Barrie, City of. 2010. *Draft Modification to the Official Plan 2009*. 25 January 2010. Accessed 30 July 2014. http://www.barrie.ca/Doing%20Business/PlanningandDevelopment/Pages/default.aspx.

Brender, N., and M. Lefebvre. 2010. *Canada's Hub Cities: A Driving Force in the National Economy*. Ottawa: Conference Board of Canada. Accessed 29 July 2014. http://www.conferenceboard.ca/e-library/abstract.aspx?did=1730.

Brewer, K. 2014. *Form without Function: Suburban Densification Trends in Dartmouth, Nova Scotia*. Master's research project, Dalhousie University. Accessed 4 August 2014. http://theoryandpractice.planning.dal.ca/suburbs/suburbs_student.html.

Brewer, K. and Grant, J.L. 2015. "Seeking Density and Mix in the Suburbs: Challenges for Mid-Sized Cities." *Planning Theory and Practice* 16 (2): 151–68. https://doi.org/10.1080/14649357.2015.1011216.

Bunce, S. 2004. "The Emergence of 'Smart Growth' Intensification in Toronto: Environment and Economy in the Official Plan." *Local Environment* 9 (2): 177–91. https://doi.org/10.1080/1354983042000199525.

Bunting, T., P. Filion, and H. Priston. 2002. "Density Gradients in Canadian Metropolitan Regions, 1971–96: Differential Patterns of Central Area and Suburban Growth and Change." *Urban Studies* 39 (13): 2531–52. https://doi.org/10.1080/0042098022000027095.

Bunting, T., P. Filion, H. Hoernig, M. Seasons, and J. Lederer. 2007. "Density, Size, Dispersion: Towards Understanding the Structural Dynamics of Mid-size Cities." *Canadian Journal of Urban Research* 16 (2): 27–52.

Calgary, City of. 1995. *Sustainable Suburbs Study*. Calgary: Planning and Development Department.

Canadian Press. 2014. "QuickList: Unemployment rates in selected Canadian cities in April," *Ottawa Citizen*, 5 August 2014.

CMHC. 2011. *Housing Market Information, Housing Now Reports*. Canada Mortgage and Housing Corporation, Ottawa. Accessed 4 September 2013. https://www.cmhc-schl.gc.ca/en/hoficlincl/homain/index.cfm

CMHC. 2013. *Housing Market Outlook, Halifax CMA*. Fall 2013. Accessed 30 July 2014. http://publications.gc.ca/collections/collection_2013/schl-cmhc/nh12-53/NH12-53-2013-2-eng.pdf.

Creese, W. 1966. *Search for Environment: The Garden City Before and After*. New Haven: Yale University Press.

Filion, P. 2009. "The Mixed Success of Nodes as a Smart Growth Planning Policy." *Environment and Planning. B, Planning & Design* 36 (3): 505–21. https://doi.org/10.1068/b33145.

Filion, P. 2010. "Reorienting Urban Development? Structural Obstruction to New Urban Forms." *International Journal of Urban and Regional Research* 34 (1): 1–19. https://doi.org/10.1111/j.1468-2427.2009.00896.x.

Filion, P., and K. McSpurren. 2007. "Smart Growth and Development Reality: The Difficult Coordination of Land Use and Transport Objectives." *Urban Studies* 44 (3): 501–23. https://doi.org/10.1080/00420980601176055.

Gonzalez, T. 2010. "Airdrie, AB: An Overview of Development Trends." Working Paper. Accessed 3 August 2014. http://theoryandpractice.planning.dal.ca/suburbs/suburbs_working.html

Gordon, D.L.A., and S. Vipond. 2005. "Gross Density and New Urbanism: Comparing Conventional and New Urbanist Suburbs in Markham, Ontario." *Journal of the American Planning Association* 71 (1): 41–54. https://doi.org/10.1080/01944360508976404.

Grant, J. 2003. "Exploring the Influence of New Urbanism in Community Planning Practice." *Journal of Architectural and Planning Research* 20 (3): 234–53.

Grant, J. 2006. *Planning the Good Community: New Urbanism in Theory and Practice. Routledge*. London, New York: Routledge.

Grant, J. 2009. "Theory and Practice in Planning the Suburbs: Challenges to Implementing New Urbanism and Smart Growth Principles." *Planning Theory & Practice* 10 (1): 11–33. https://doi.org/10.1080/14649350802661683.

Grant, J., and A. Curran. 2007. "Privatized Suburbia: The Planning Implications of Private Roads." *Environment and Planning. B, Planning & Design* 34 (4): 740–54. https://doi.org/10.1068/b32136.

Grant, J.L., and K. Perrott. 2009. "Producing Diversity in a New Urbanism Community: Policy and Practice." *Town Planning Review* 80 (3): 267–89. https://doi.org/10.3828/tpr.80.3.3.

Grant, J.L., and K. Perrott. 2011. "Where Is the Café? The Challenge of Making Retail Uses Viable in Mixed Use Suburban Developments." *Urban Studies* 48 (1): 177–95. https://doi.org/10.1177/0042098009360232.

Hanna, K.S. 1997. "Regulation and Land-use Conservation: A Case Study of the British Columbia Agricultural Land Reserve." *Journal of Soil and Water Conservation* 52 (3): 166–70.

Harris, R. 2004. *Creeping Conformity: How Canada Became Suburban, 1900–1960*. Toronto: University of Toronto Press.

Howard, E. 1985 [1902]. *Garden Cities of To-morrow*. Ill. repr. ed. Eastbourne, UK: Attic Books.

HRM (Halifax Regional Municipality). 2006. *Halifax Regional Municipality Planning Regional Planning Strategy*. Halifax: Planning and Development Services, Halifax Regional Municipality.

Isin, E., and R. Tomalty. 1993. *Resettling Cities: Canadian Residential Intensification Initiatives*. Ottawa: Canada Mortgage and Housing Corporation.

Katz, Peter, ed. 1993. *The New Urbanism: Towards an Architecture of Community*. New York: McGraw Hill.

Langley, Township of. 2014. *Overview: Township of Langley*. Accessed 4 August 2014. https://www.tol.ca/About-the-Township/Overview.

Langlois, P. 2010. "Municipal Visions, Market Realities: Does Planning Guide Residential Development?" *Environment & Planning B* 37 (3): 449–62. https://doi.org/10.1068/b34103.

Leffers, D., and P. Ballamingie. 2013. "Governmentality, Environmental Subjectivity, and Urban Intensification." *Local Environment* 18 (2): 134–51. https://doi.org/10.1080/13549839.2012.719016.

Lehrer, U., and R. Milgrom. 1996. "New (Sub) Urbanism: Countersprawl or Repackaging the Product." *Capitalism and Socialism* 7 (2): 49–64. https://doi.org/10.1080/10455759609358678.

Leo, C., and W. Brown. 2000. "Slow Growth and Urban Development Policy." *Journal of Urban Affairs* 22 (2): 193–213. https://doi.org/10.1111/0735-2166.00050.

Lepawsky, J., H. Hall, and B. Donald. 2014. "Kingston and St John's: The Role of Relative Location in Talent Attraction and Retention." In *Seeking Talent for Creative Cities: The Social Dynamics of Innovation*, ed. J.L. Grant, 201–18. Toronto: University of Toronto Press.

Maoh, H., M. Koronios, and P. Kanaroglou. 2010. "Exploring the Land Development Process and Its Impact on Urban Form in Hamilton, Ontario." *Canadian Geographer / Le Géographe canadien* 54 (1): 68–86. https://doi.org/10.1111/j.1541-0064.2009.00303.x.

Millward, H. 2002. "Peri-urban Residential Development in the Halifax Region, 1960–2000: Magnets, Constraints, and Planning Policies." *Canadian Geographer / Le Géographe canadien* 46 (1): 33–47. https://doi.org/10.1111/j.1541-0064.2002.tb00729.x.

O'Neill, D. 2000. *The Smart Growth Tool Kit: Community Profiles and Case Studies to Advance Smart Growth Practices*. Washington, DC: Urban Land Institute.

Ontario, Province of. 2012. *Simcoe Sub-area Amendment, January 2012*. Places to Grow. Ministry of Economic Development, Employment and Infrastructure. Accessed 30 July 2014. https://www.placestogrow.ca/index.php?option=com_content&task=view&id=210&Itemid=15

Perrin, L., and J.L. Grant. 2014. "Perspectives on Mixing Housing Types in the Suburbs." *Town Planning Review* 85 (3): 363–85. https://doi.org/10.3828/tpr.2014.22.

Perrott, K. 2008. "Markham, ON: An Overview of Development Trends." Working Paper. Accessed 3 August 2014. http://theoryandpractice.planning.dal.ca/suburbs/suburbs_working.html.

Punter, J. 2003. *The Vancouver Achievement: Urban Planning and Design*. Vancouver: UBC Press.

Quastel, N., M. Moos, and N. Lynch. 2012. "Sustainability-as-Density and the Return of the Social: The Case of Vancouver, British Columbia." *Urban Geography* 33 (7): 1055–84. https://doi.org/10.2747/0272-3638.33.7.1055.

Rosen, G., and K. Brewer. 2013. "Barrie at a Crossroad: Dilemma of a Mid-sized City." Working Paper. Accessed 4 August 2014. http://theoryandpractice.planning.dal.ca/suburbs/suburbs_working.html.

Scott, D. 2010. "Township of Langley, BC: An Overview of Development Trends." Working Paper. Accessed 3 August 2014. http://theoryandpractice.planning.dal.ca/suburbs/suburbs_working.html.

Searle, G., and P. Filion. 2011. "Planning Context and Urban Intensification Outcomes: Sydney versus Toronto." *Urban Studies* 48 (7): 1419–38. https://doi.org/10.1177/0042098010375995.

Sewell, J. 2009. *The Shape of the Suburbs: Understanding Toronto's Sprawl*. Toronto: University of Toronto Press.

Sinoski, K. 2013. "The New Surrey: Developing Six Cities at Once." *Vancouver Sun*, 26 January. Accessed 3 August 2014. http://www.vancouversun.com/Surrey+Developing+cities+once/7874086/story.html.

Skaburskis, A. 2006. "New Urbanism and Sprawl: A Toronto Case Study." *Journal of Planning Education and Research* 25 (3): 233–48. https://doi.org/10.1177/0739456X05278985.

Statistics Canada. 2008. *Community Profiles, 2006 Census*. Statistics Canada Catalogue No. 92–591-XWE. Ottawa: Statistics Canada. Accessed 29 July 2014. http://www5.statcan.gc.ca/olc-cel/olc.action?objId=92-591-X&objType=2&lang=en&limit=0.

Statistics Canada. 2013. *NHS Focus on Geography Series*. 99–010–X2011005. Ottawa: Statistics Canada. Accessed 29 July 2014. http://www12.statcan.gc.ca/nhs-enm/2011/as-sa/fogs-spg/?Lang=E.

Statistics Canada. 2014. *Population and Dwelling Counts, for Census Metropolitan Areas, 2011 and 2006 Censuses*. Ottawa: Statistics Canada. Accessed 29 July 2014. http://www12.statcan.gc.ca/census-recensement/2011/dp-pd/hlt-fst/pd-pl/Table-Tableau.cfm?T=205&S=3&RPP=50.

Taylor, Z., M. Burchfield, and A. Kramer. 2014. *Alberta Cities at the Crossroads: Urban Development Challenges and Opportunities in Historical and Comparative Perspective*. SPP Research Paper No. 7. Accessed 7 August 2014. http://papers.ssrn.com/sol3/papers.cfm?abstract_id=2474377 https://doi.org/10.2139/ssrn.2474377.

chapter nine

Latin America: The Suburb Is the City

ALAN GILBERT

Until the twentieth century most Latin American cities were tiny. The median population of the forty largest cities in 1880 was about 35,000 and only eight cities in Latin America had a population in excess of 100,000 (Sargent 1993). Urban form tended to follow a common pattern because the Spanish and, to a lesser extent the Portuguese, laid down firm rules for urban development in their American colonies. "Though colonial cites differed widely in functions and origins – encompassing what might originally have been missions, garrisons, administrative centres, local or regional markets, mining camps, ocean-side trading posts, indigenous centres and shrines – development typically followed the original city plans, which appealed to the enlightened successors of the early colonial elite" (Fernandez-Armesto 2013, 371). The new cities were planned on a gridiron basis that spread out from the administrative and ecclesiastical buildings located around the main square. Originally, the most advantaged social groups in the society lived in areas of easy access to the seats of governmental and ecclesiastical power that surrounded the central plaza. Less advantaged groups tended to occupy more peripheral locations.

Most people, and particularly the rich, lived within this planned area until at least the late nineteenth century. Amato (1970, 97) notes: "From the founding of the city in the sixteenth century until the end of the nineteenth century, the wealthy of Bogotá lived in several barrios surrounding the Plaza de Armas." Similarly, in Quito, "the upper income groups lived around the city's major plaza from the early founding of the city until the first few decades of the present century" (ibid., 98). Given their limited populations and the fact that poor transport facilities meant virtually everyone had to walk to work, there was little need for suburbs.

This situation changed dramatically during the twentieth century, when Latin American cities were transformed by demographic growth, technological change and government policy. Suburban expansion formed a key part in the region's urban development. São Paulo's urbanized area increased five times between 1914 and 1930 and nine times during the following fifty years (Bonduki 1983; Gilbert 1996, 90). Similarly, Bogotá's territory increased fourteen times between 1938 and 2010 (Gilbert 1978, 91; SDP 2010) and that of Lima twenty-six times between 1940 and 1990 (Gilbert 1996, 90).

The Causes of Suburban Expansion

Demographic Growth

Urban growth in Latin America during the second half of the twentieth century was much faster than it had ever been in Europe or North America. Between 1880 and 1914 there was an early spurt of urban expansion in Argentina, Uruguay and Southern Brazil, fuelled by immigration from Italy, Portugal and Spain. Elsewhere urban expansion occurred later and was predominantly composed of internal migrants. Between 1950 and 2011 the urban population of Latin America and the Caribbean increased from around 69 million to 472 million. The impact of migration on Latin America was profound. In 1940, one-third of the population lived in cities, one-half twenty years later and two-thirds by 1980. In absolute terms, the figures are still more astounding. After 1940, most of the larger cities were growing rapidly, some doubling and even tripling their population in twenty years. During the 1940s, Cali (Colombia), Caracas (Venezuela) and São Paulo all grew annually by more than 7 per cent. By 1980, eighteen cities had more than 1 million inhabitants and today six cities have achieved so-called mega-city status (Table 9.1).

At first, most of the new arrivals were accommodated in the central areas, renting accommodation in some form of tenement (Gilbert and Varley 1991). But the influx of people was so great that alternatives were soon needed. The vast bulk of this growth was necessarily absorbed in areas beyond the original city hubs. As such, the great majority of Latin America city dwellers soon began to live in the suburbs.

Table 9.1. Latin America's largest urban agglomerations, 1950–2010 (population in millions)

City	1950	1980	2010
São Paulo (Brazil)	2.334	12.089	19.649
Mexico City (Mexico)	2.883	13.010	20.142
Buenos Aires (Argentina)	5.098	9.422	13.370
Rio de Janeiro (Brazil)	2.950	8.583	11.867
Lima (Peru)	1.066	4.438	8.950
Bogotá (Colombia)	630	3.525	7,353
Belo Horizonte (Brazil)	412	2.441	5.407
Santiago (Chile)	1.322	3.721	5.949
Guadalajara (Mexico)	403	2.269	4.442
Porto Alegre (Brazil)	488	2.133	3.892
Brasília (Brazil)	36	1.293	3.701
Monterrey (Mexico)	356	1.992	4.100
Recife (Brazil)	661	2.122	3.684
Salvador (Brazil)	403	1.683	3.947
Fortaleza (Brazil)	264	1.488	3.520
Medellín (Colombia)	376	1.731	3.595
Curitiba (Brazil)	158	1.310	3.118
Caracas (Venezuela)	694	2.575	3.176

Source: UNDESA (2011)

Public Transport

Suburbia can develop only when public transport allows people to travel to work. In Latin America, the beginnings of suburban growth came with the emergence of railways and trams and the ability this gave the affluent classes to move out of the city centre into more environmentally desirable areas (Amato 1970). In Mexico City, the first tram service was introduced in the late 1850s and an extensive network of routes was in operation by the 1880s (Morrison n.d.). In Buenos Aires, the first private tram routes appeared in the 1860s (Scobie 1964). In Bogotá the first tramway was built in 1886. None of these systems were electrified until the early 1900s. In São Paulo, the opening of the railway to the port of Santos in 1867 was the first stimulus to metropolitan growth (Meyer, Grostein, and Biderman 2004, 35).

The emergence of public transport led to the growth of suburbs for the elite. The rich moved to beach areas like Copacabana in Rio de Janeiro and Miraflores in Lima, to the green hills in the north of Bogotá and Quito and to the lower, less inhospitable, slopes of that most elevated of national capitals, La Paz. In Santiago, Chile, a "major move of high income groups occurred in the areas of Providencia during the 1920s and has been followed by continued pressure over the past several decades within the same general sectors spilling into the areas of Las Condes and Nuñúa" (Amato 1970, 101).

However, life for the majority hardly changed because few effective public transport systems were operating before 1940. Horse-drawn trams were still operating in Rio de Janeiro in 1926 and there were still only 550 electric trams operating in São Paulo as late as 1934. Real progress was confined to Buenos Aires, where the electric tram was operating over a 405-mile system and was carrying 3,245,000 passengers a year in 1919 (Scobie 1974). Falling fares allowed some workers to join the middle class in the move to the suburbs.

Elsewhere, the growth of suburbia awaited the coming of the bus – typically a phenomenon of the 1930s and 1940s. The bus transformed the nature of poor people's lives. Now that they could commute longer distances, increasing numbers moved out of the inner-city tenements into the new shanty towns. In São Paulo "the bus lines made it possible to link up even the most distant residential areas with the workplace. This in turn led to intense property speculation, as unoccupied land was transformed into plots to be sold off to the droves of workers coming in the wake of industrial expansion" (Kowarick and Ant 1988, 10).

Better transportation led to a significant fall in residential densities. The "urbanized area density" of Buenos Aires fell dramatically from 207 persons per hectare in 1890 to 60 thirty years later (Angel et al. 2010). Similarly, in Mexico City the density fell from 675 in 1920 to 270 twenty years later and continued to fall ever after. In Santiago, the density fell from 184 in 1910 to 112 in 1930. What is clear from these figures is that urban densities in Buenos Aires, Santiago and São Paulo were never as high as they were in Guatemala City or Mexico City. Population densities in the region's cities continue to be very different (see below).

In recent years, vast sums have been invested in improving transport infrastructure to allow people to get to work, go to school, and escape the city at the weekend. Today, most of the region's larger cities contain extensive road systems, freeways, bus terminals, railway networks, and metros. Cities have expanded physically and most have reaped the downside of modernity: traffic congestion, road accidents, and air pollution. The number of vehicles has risen, often dramatically. Mexico's Federal District and its neighbouring state saw an increase in the number of cars from 2.28 million in 1980 to 5.23 million in 2006. Latin America now has some 60 million cars, most of which are concentrated in the cities.

Government Policy

Governments played a complicated role in the urban growth process. They provided services, they developed master plans, they built social housing, they encouraged private banks to provide credit for homeownership, and increasingly they provided subsidies to reduce the cost of transport, services, and housing. Of course, given their limited resources and the poverty of so many of the inhabitants, few governments were able to help the majority of poor people significantly. In any case the distribution of most of the subsidies and resources was slanted towards the better off.

Rapid suburban growth was also dependent on extensive investment in public utilities. Modern engineering provided the means to generate and distribute electricity and water. Industrial and commercial development generated the taxes to develop education and health systems. However deficient the servicing of the low-income areas, something was done to placate their inhabitants and to prevent the spread of epidemics. Since 1945 major improvements have taken place in most Latin American cities, much of it financed through loans from the World Bank and the Inter-American Development Bank. The bulk of the population in major cities had access to electricity and potable water by 1980; by 1990 supply was almost universal, albeit not without its problems (Table 9.2).

A further factor encouraging suburban development in the region was a misguided passion for both home and private car ownership. Every government bar one bought into the mythologized ideology of the United States and pretended that everyone could be a homeowner and even that exception, Castro's Cuba, encouraged homeownership (Gilbert 2016). Homeownership went hand in hand with suburban growth and,

Table 9.2. Urban homes with electricity, water, and sewerage provisions in selected Latin American countries, 1990–2011 (percentage)

Country	Year	Piped water	Sewage disposal	Electricity
Argentina	1990	97	58 (1995)	100
	2011	99	67	100
Chile	1990	97	84	99
	2010	99	98	99
Colombia	1990	98	95	99
	2006	98	94	100
Guatemala	1990	87	70	87
	2006	94	66	96
Mexico	1990	94	79	99
	2011	97	91	99
Uruguay	1990	95	57	98
	2011	97	65	100

Source: UNECLAC (2007, 71) and UNECLAC (2012, Table 1.5.1)

along with growing prosperity and rising levels of private car ownership, it led inevitably to urban sprawl. Even the poor participated in the suburban dream, albeit through an inferior, self-help version.

Incremental Housing

In 1940 few families owned their homes and the vast majority were tenants. Many rented rooms in colonial-style buildings close to the city centre. A typical house had two storeys, the rooms overlooking a central patio that contained services such as toilets and washing facilities. The precise architectural details differed between cities and between social classes, but the basic design was similar. Every income group lived in this kind of accommodation: affluent families occupied the whole house, together with their large complement of domestic servants; poor families shared the house, each household occupying a single room.

Incipient self-help housing areas appeared in some cities in the late nineteenth century, but they emerged on a large scale only when public transport improved (de Ramón 1985; Scobie 1974). In Bogotá the residential map of 1927 shows few areas of informal housing and, in

Caracas, few *barrios* had developed before 1930 (Morris 1978). Lima had only five *barriadas* in 1940 (Matos Mar 1968). Table 9.3 illustrates how informal shelter became an increasingly important element in housing the poor. However dubious the data and the definitions that underpin them, the trend is clear.

Today, of course, millions of people live in self-help housing and more than half of the population of many Latin American cities live in housing that began informally – without either planning permission or services and sometimes with ambiguous property rights. Governments rarely opposed this kind of development and sometimes encouraged it, because it represented an essential social safety valve. Given rapid population growth, generalized poverty, the highly skewed distribution of wealth, a land market that was often controlled by private land monopolies and a building industry geared to formal construction methods, a "proper" house was clearly beyond the reach of most poor families.

Cities in the region differed, however, in the extent to which they could offer homeownership even of the self-help kind. In cities surrounded by cheap land, for instance, Lima, with its desert margins, and Rio and Caracas with their steep hillsides, self-help expansion was easy and for many years governments turned a blind eye to land invasions (Collier 1976; Ray 1969). However, where peripheral land was fertile, its often powerful owners could persuade the authorities to police the land and prevent its invasion. In cities like Bogotá and Quito, self-help settlers were channelled towards illegal subdivisions, where they were required to buy their self-help plots (Doebele 1975; Gilbert 1981). In such cities many families lacked the resources with which to buy a plot and continued to rent accommodation. To this day, around half of the population in those cities still rent their homes, increasingly in the consolidated self-help suburbs. Self-help suburban settlers have progressively become landlords (Blanco, Gilbert, and Kim 2016).

Over the years government approaches to housing and urban development have changed, radically affecting the nature of suburban growth. In the 1960s and 1970s military dictatorships often destroyed invasion settlements, particularly those occupying land close to elite settlements. Increasingly, however, international policies began to recognize that self-help housing was inevitable and to recommend policies to upgrade and service the so-called marginal settlements. Property titles were distributed widely, based on the mythology that a piece of paper would bring prosperity to the poor; banks would then lend, businesses would flourish and poverty would disappear (de Soto 2000;

Table 9.3. The growth of self-help housing in selected Latin American cities, c.1940–c.1990

City	Year	Population (000s)	Self-help population (000s)	Self-help population/total (per cent)
Buenos Aires	1956	6,054	109	2
	1970	8,353	434	5
	1980	9,766	957	10
Caracas	1961	1,330	280	21
	1964	1,590	556	35
	1971	2,200	867	39
	1985	2,742	1,673	61
Lima	1956	1,397	112	8
	1961	1,846	347	17
	1969	3,003	805	24
	1981	4,601	1,150	25
	1991	4,805	1,778	37
Mexico City	1952	2,372	330	14
	1966	3,287	1,500	46
	1970	7,314	3,438	47
	1976	11,312	5,656	50
	1990	15,783	9,470	60
Rio de Janeiro*	1947	2,050	400	20
	1957	2,940	650	22
	1961	3,326	900	27
	1970	4,252	1,276	30
	1990	5,366	1,076	20

* Data for the municipality rather than for the metropolitan area, which underestimates the informal population

Source: Adapted from Gilbert (1998). Rio de Janeiro 1990 data from Xavier and Magalhães (2003)

Gilbert 2002b). Later, Chilean-type policies tried to slow informal settlement offering subsidies to the poor and encouraging private builders to construct formal social housing on cheap peripheral land.

Unfortunately, few governments managed to plan for the growth of new self-help settlements. Sites and service schemes appeared sporadically in the 1970s and then disappeared from the face of the earth. The poor continued to settle in the increasingly distant periphery, too often occupying areas at risk from flood, hurricane and landslide. Latin America's governments have never been able to get ahead of the informal settlement curve, even though the pace of demographic growth has slowed throughout the region.

Decentralization of Employment

Modern industry emerged towards the end of the nineteenth century in the more prosperous cities of Argentina, southern Brazil and Chile. Industrial plants were set up close to port areas or along major road and rail routes on the fringe of the city. In 1950 most of São Paulo's industry was located along a 35-kilometre strip close to the main rail routes (Meyer, Grostein, and Biderman 2004, 166). When industrialization accelerated throughout the region after 1940, most cities developed specialized industrial areas and self-help housing sprouted close to these centres of employment.

Later, growing traffic congestion and rising land prices encouraged the de-concentration of office employment, which led to further territorial expansion. Over time this process produced polycentric development – sub-centres of employment appearing in most of the largest cities (Aguilar and Alvarado 2005; Suárez and Delgado 2009; Meyer, Grostein, and Biderman 2004).

In time shopping malls and hypermarkets began to penetrate the urban fabric. These retail centres both encouraged and followed residential development. They helped introduce a new shopping culture and style of living to Latin America and encouraged further use of the automobile. Nowhere is this pattern better exemplified than in the Barra da Tijuca area of Rio de Janeiro. This high-income neighbourhood developed along the Avenida das Américas once a tunnel was opened to connect Barra with the existing elite areas of Rio in the late 1960s. During the 1980s the area grew rapidly as middle- and upper-class *cariocas* moved in to take advantage of the shopping malls, theatres, universities and, of course, the 17 kilometres of beach. The foreign flavour of the area is summed up in the names of the many shopping malls: BarraShopping, the New York City Center, Downtown and Barra Square. Walmart, Carrefour and Bon Marché all have supermarkets in the area.

Diversity of Suburban Form

There is no universally acceptable definition of a suburb. Vaughan et al. (2009, 475) argue: "Beyond the most perfunctory level of definition, it is far from clear as to what this term actually means or indeed, whether it can be thought to possess meaning at all. A pronounced tendency to neologism in suburban studies highlights the underlying theoretical weakness." Similarly, Hinchcliffe (2005, 899) states: "The literature on suburbs is extensive, yet the subject always seems elusive. For some, the suburb is a geographical space; for others, a cultural form; while for others still, it is a state of mind."

The problem with defining the suburb is that, because most cities expanded so rapidly in the late nineteenth and the twentieth centuries, the great majority of people now live in suburbs. And, given the diversity of incomes and the time over which it has evolved, many different kinds of suburbia have emerged. In fact, Jauhiainen (2013) identifies nine types of suburbia: terraced suburbs, villa suburbs, industrial and working-class suburbs, garden suburbs, extended suburbs, gated communities, squatter and shanty-town suburban areas, suburban sprawl and suburban edge cities.

Not all of these sub-types are found in Latin America, but its suburbs still form a fascinating mosaic that has emerged within an urban framework of wealth and poverty. They range from affluent low-density, American-style suburbs to impoverished areas of subsidized social housing. The informal suburbs consist of a range of settlement from flimsy unserviced shacks to consolidated high-density suburbs containing a majority of tenants.

Elite Suburbs

The equivalent of "villa suburbs" emerged around most Latin American cities as soon as transport facilities began to improve (Abreu 1987; Amato 1970; Marques and Bitar 2002, 130). These suburbs often displayed a mixture of housing designs, as each home had its own architect. However, particular styles became fashionable. In Bogotá one of the early suburbs, Teusaquillo, was dominated by so-called English-style architecture. By the 1950s California-style design and, later, post-modern designs were sweeping through the region. As cities expanded, elite barrios tended to develop further and further from the central city. In Rio the fashionable suburbs moved ever outwards along the southern beaches; Gloria, Flamengo, Botafogo, Urca, Copacabana, Ipanema, Leblon, São Conrado and, most recently, Barra da Tijuca.

As urban populations became more prosperous, middle-class estates emerged that mimicked the styles of the higher-income areas. Over time, however, these developments began to adopt more standard designs and look more and more like suburbs anywhere across the world.

Social Housing

From the 1920s on most Latin American governments began to construct houses for the poor, or at least for the more powerful unionized groups among them (the military, police, dockers and power workers). The Banco Obrero began operations in Venezuela in 1928 and the Institute for Territorial Credit was set up in Colombia in 1939. Such agencies became particularly important during the period of the Alliance for Progress (1961–9), when USAID and the Inter-American Development Bank poured money into the region, a substantial chunk of it for housing. Many Latin American cities bear the strong mark of government building programs – Ciudad Kennedy in Bogotá and Villa Isabelica in Valencia – although the most widespread development of this approach was in the new "planned" cities such as Brasília and Ciudad Guayana. Some of this housing was sensitively designed and constructed, although the quality generally deteriorated as the years went by as governments attempted to cut costs. Nonetheless, because of the excess of housing demand over supply, long queues developed and official allocation systems were often bypassed or corrupted. As a result, few really poor people obtained subsidized units (Klak 1992; Laun 1977; Mayo 1999). These estates tended to be built on the fringes of the city, where land prices were lowest, thereby contributing further to urban sprawl.

Public housing policy changed dramatically in the 1970s, commencing in Chile (Almarza 1997; Arellano 1982; Gilbert 2002a). Public housing would no longer be contracted by the state but would be built by the private sector, in response to market signals. Instead of the government specifying what the private sector should produce, builders would compete in providing consumers with the kind of housing that they wanted. The assumption was that private enterprise would produce more cheaply than under the public contracting system and would provide a wider choice of housing for the poor.

To stimulate demand, housing subsidies would be given to poor families, who would buy their homes from private developers. The Chilean government devised a way of allocating subsidies to families who were both poor and prepared to help themselves. The test of the latter was their willingness to accumulate savings: the longer their savings record and the greater their savings, the more likely they were to get a subsidy. The subsidy and savings would not cover the whole cost of the house and the difference would be covered by a mortgage. After initial doubts, Washington embraced the Chilean model because it fitted the World Bank's new policy approach: the need to completely reorganize how housing in poor countries was financed and administered. The Chilean model embraced three admirable elements: explicit targeting of the poor, transparency and private market provision. Subsidies were acceptable to the World Bank provided they were limited in number, aimed at the poor and would help to stimulate demand.

By 1993 the Chilean model had become acknowledged "best practice" in the region by the World Bank, the Inter-American Development Bank and USAID, which encouraged its diffusion (Gilbert 2002a; Kimm 1993). Eventually, the ABC package (*ahorro, bono y crédito*) was adopted in various forms in numerous Latin American countries, including Colombia, Costa Rica, Ecuador, Mexico, Panama and Peru (Held 2000; Pérez-Iñigo González 1999). Subsequently, other countries such as Brazil and Mexico have taken up some variation of the model (Jha 2007). This approach has produced increasing numbers of houses and led to the development of new housing estates sometimes miles from the central area.

Despite its widespread popularity among governments and the almost certain superiority of the ABC mechanism compared with earlier approaches, major question marks have been placed against its success (Gilbert 2004; Giraldo 1994, 1997; Ducci 1997; Rojas and Greene 1995; Klaufus 2010). One major criticism has been its propensity for increasing urban sprawl (see below).

High-Rise Suburbs

As cities grew in size, increasing traffic congestion, rising land prices and fear of crime often escalated and more people began to live in high-rise apartments. This trend was particularly marked in cities located in narrow valleys or whose growth was constrained by mountains or the sea. By the 1970s many elite families in Bogotá, Caracas, Rio de Janeiro and São Paulo lived in this kind of accommodation and gradually increasing numbers of middle-class families did so too. The result was that in Bogotá, 50.6 per cent of families lived in apartments in 2005 compared with only 29 per cent in 1964. And, of the 130 buildings with more than fourteen floors in 2013, only 25 had been inaugurated before 1970, of which only 13 were for housing (Wikipedia 2013). By contrast, 49 high-rise buildings of fourteen floors or more were inaugurated between 2000 and 2013, of which 38 were apartment blocks.

Squatter and Shanty Towns

Few Latin Americans are homeless, because many families build or organize the construction of their own homes. Ordinary families scrape together the funds to buy the bricks and cement, the glass and window frames. If the authorities eventually provide services, help seldom comes when the poor first need it.

Most incremental housing in Peru, the Caribbean coasts of Colombia and Venezuela, parts of Brazil and much of Central America has been constructed on land occupied through invasion. But where governments have discouraged this process, the poor have tended to buy plots in pirate or clandestine developments. These settlements lack infrastructure, services and properly registered title deeds, but they are illegal only in a technical sense.

Few of the illegal forms of land development are under threat of removal by the authorities. The only settlements threatened are those on land scheduled for prestige projects or close to high-income residential areas. Once families feel that they have a certain security of tenure, they improve their dwellings and most Latin American cities now have older self-help settlements containing a majority of houses with several storeys. Over time, the authorities bring in services and in the better-managed cities the great majority of homes have formal electricity, water and sewerage provision.

Of course, self-help construction has its problems. Many homes are constructed on land susceptible to flood or landslide. Some buildings may not survive earthquakes or hurricanes. Self-help housing has contributed to urban sprawl and added to journey times and traffic congestion. Critically, poverty has led to people living for years in uncomfortable and sometimes unsafe shelter. While self-help housing has accommodated people who might have been homeless, it has not been a wholesale success and, particularly in badly managed cities, many settlements continue to lack proper services and infrastructure for years.

Gated Communities

Some form of gated community has been around for a long time in the form of golf and riding clubs. In Mexico City these associations first appeared in the 1920s and contained a limited number of peripheral residences (García Peralta and Hofer 2006). Later, they were transformed into real-estate businesses, providing areas for the sale of plots of land for residential development. In the 1970s rising burglary and armed-assault rates in inner-city residential areas led to a perception of insecurity – albeit one that was often exaggerated – and the building industry began to construct high-security enclaves in enclosed developments for luxury housing. This produced a shift among wealthy families to the outskirts and changed the structure of traditional residential areas in the city. The most spectacular of these gated cities is the Santa Fe district, which lies 18 kilometres southwest of the Zócalo, the city's main square.

In São Paulo the first low-density, high-income suburb, Chácara Flora, appeared in the 1920s and this kind of development became a metropolitan phenomenon during the construction boom of the late 1960s (Marques and Bitar 2002, 130). Alphaville, developed in the mid-1970s, was the first low-rise *condomínio fechada*. Since then many more have appeared, particularly on the western edge of the city. The first vertical gated community, Ilha do Sol, opened in 1973 (Caldeira 2000).

Gradually, less luxurious developments appeared to satisfy the increasing middle-class demand for such accommodation. In the more mixed middle- and upper-income, inner-city residential areas, these techniques were adopted in milder forms. The population began to close off neighbourhoods and streets and to protect the entrances with gates and guards. Anti-burglary systems were adapted to protect the traditional courthouses, while windowless facades and massive garage gates turned the houses into fortresses surrounded by an ocean of perceived insecurity.

As in the United States, more and more gated communities are emerging. In Buenos Aires, Santiago, São Paulo and Rio de Janeiro the rich increasingly live in closed estates with guards keeping out all but employees and invited guests. Some gated communities have everything from offices and supermarkets to recreation facilities. Residents feel safer from crime, fear of which is one of the key reasons why these communities are proliferating.

The Consequences

Urban Sprawl

Mills and Tan (1980, 320) praise: "Continued urban decentralisation in developing countries [because] actual densities near centres of large urban areas in developing countries are extremely high. It is hardly surprising that modest income increases induce people to move out." They also claim that "flattening urban population density functions must be among the most pervasive and best-documented trends in the developed world, extending throughout the period of rapid industrialisation and urbanisation. There appears to be no country in which the trend has not been pervasive and persistent during many decades" (ibid., 316), something that Angel et al. (2010) have recently confirmed with their rule of thumb for planning – when the population of a city doubles, its area triples.

Nevertheless, urban sprawl has been widely condemned. UN-HABITAT (2010) argues: "Urban sprawl has a negative impact on infrastructure and the sustainability of cities. In most cases, sprawl translates to an increase in the cost of transport, public infrastructure and of residential and commercial development. Moreover, sprawling metropolitan areas require more energy, metal, concrete and asphalt than do compact cities because homes, offices and utilities are set farther apart."

Rising levels of car ownership have created external diseconomies; traffic jams are increasingly common, air and water pollution is affecting more and more cities and too many urban areas are spreading into administrative jurisdictions where the authorities are incapable of properly servicing and organizing the consequent urban sprawl. The failure of most governments to raise enough taxes or to benefit from escalating land values by levying betterment charges has held down revenues and prevented them from redistributing resources.

Most cities in Latin America have succumbed to the problem of urban sprawl (Monkkonen 2011; Janoschka 2002; Abramo 2009; De Mattos 1999) and there is plenty of evidence that car-based, low-density residential development has been occurring in increasingly distant areas. In Bogotá, municipalities up to 30 kilometres from the city centre, such as Chía, La Calera and Sopó, are attracting increasing numbers of well-off families.

Few governments have attempted to control the growth in car ownership and planning regulations and servicing policies have been permissive with respect to urban expansion. The rapid growth of middle-class and informal suburbs has created traffic congestion and raised the cost of providing infrastructure. In addition, government social housing programs have contributed to urban sprawl. In Chile, the capital subsidy program "is one of the major causes of the uncontrolled spread of the Chilean capital" (Paquette-Vassalli 1998, 369). The rapid expansion of subsidized housing and/or cheap mortgage credit in Brazil and Mexico has had a similar effect. Too much of the new housing has been located far from the existing urban areas and has not been integrated into the transport of infrastructure networks. In Mexico, cheap industrialized and homogeneous housing designs have failed "to make a city" (Iracheta 2011, 98). The houses have been located far from the city and lack adequate transport links, water-treatment facilities and parks (García Peralta and Hofer 2006). There is increasing concern about the large number of homes that are unoccupied and even abandoned (Puebla 2011). Bonduki (2011, 94) makes a similar complaint about Brazil, recognizing that official policy to increase homeownership and access to credit has not changed the existing urban model.

Such low-density, urban development is reflected in falling population densities in some cities. Angel et al.'s (2010) figures show significant falls in urbanized area densities for Guatemala City and Mexico City between 1950 and 2000. However, the figures for Buenos Aires, Santiago and São Paulo actually show increases during much of that period. Figures for Bogotá are similar: urban density has increased over time. While the urban area increased from 2,514 hectares in 1938 to 34,000 in 2010, this 13.5-fold increase was much less than the 21-fold growth in population (Gilbert 1978, 91; SDP 2010). The result is that some parts of the self-help periphery exhibit much higher population densities than the central area or indeed most other areas of the city.

However, Latin American cities differ considerably and Table 9.4 shows the great diversity in their population densities. While all are much more densely populated than their North American neighbours, Bogotá and Lima figure among the most densely populated cities of

Table 9.4. Latin American cities ranked by population density

World ranking	City/urban area	Country	Population	Land area (km²)	Density (person per km²)
9	Bogotá	Colombia	7,000,000	518	13,500
11	Lima	Peru	7,000,000	596	11,750
25	São Paulo	Brazil	17,700,000	1,968	9,000
27	Mexico City	Mexico	17,400,000	2,072	8,400
28	Santiago	Chile	5,425,000	648	8,400
31	Recife	Brazil	3,025,000	376	8,050
35	Rio de Janeiro	Brazil	10,800,000	1,580	6,850
36	Monterrey	Mexico	3,200,000	479	6,700
39	Guadalajara	Mexico	3,500,000	596	5,900
46	Buenos Aires	Argentina	11,200,000	2,266	4,950
49	Porto Alegre	Brazil	2,800,000	583	4,800
51	Belo Horizonte	Brazil	4,000,000	868	4,600
52	Fortaleza	Brazil	2,650,000	583	4,550
75	Quito	Ecuador	1,500,000	479	3,150
86	Brasilia	Brazil	1,625,000	583	2,800

Source: Adapted from http://www.citymayors.com/statistics/largest-cities-density-125.html

the world, whereas Brasília, given its distinctive and much discussed urban design, exhibits a much lower density.

The patterns are clearly inconsistent. In some cities, population densities have fallen spectacularly, whereas in others urban sprawl has gone hand in hand with increasing population densities. The most obvious explanation for the latter tendency is that the growing distance between home and work, when allied to severe traffic congestion, is discouraging some people from moving ever outwards. The expansion of high-rise living is creating more opportunities for the middle classes to live securely and closer in.

In the case of the poor, another mechanism may be operating in some cities. Access to cheap land has become more difficult and fewer people can obtain a plot on which to build their own home. This scarcity has led to a rise in the numbers of tenants, particularly in the old and now consolidated self-help settlements. In such areas Bogotá tenants now outnumber owners, the former self-help builders having created space for later generations (Gilbert 1999).

Social Segregation

Latin American cities have long been divided according to income, ethnicity, and social class. Fortunately, the gulf that once divided the urban oligarch and the new migrant from the countryside is smaller today because more and more urbanites have been born in the cities and received at least some education. In addition, as countries have grown richer, the middle class has grown rapidly. Many more people work in offices and live in some kind of formal housing and own their own car. In the consolidating self-help areas, many people are increasingly adopting middle-class aspirations and even modest homes are replete with a range of consumer durables.

Nevertheless, UN-HABITAT (2010) argues that "urban sprawl adds to the urban divide, pushing social segregation along economic lines that result in spatial difference in wealth and quality of life across various parts of cities and metropolitan areas, run down inner cities and more suburbs." Certainly, urban sprawl containing housing estates with homes of similar price means that different social classes now often live a long way apart.

Others argue that residential segregation is increasing because of the growing number of gated communities, with their "walled condominiums, apartment buildings guarded by security towers, private policing, 'armed response,' and so on" (Sa 2007, 154). Some of these communities are virtually autonomous. With their own offices, supermarkets and recreational facilities, some residents rarely need to leave. The appeal of these estates is that their inhabitants feel they are safe from crime and from the more unsavoury aspects of urban life. The walls and gates that surround the communities "prevent people from seeing, meeting and hearing each other; at the extremes, they insulate and they exclude" (Marcuse and van Kempen 2000, 250).

Whether gated communities cut the rich off more from ordinary people than before is debatable. The old colonial-style house was impenetrable except through the main door and the high-rise apartments that proliferated throughout the region as land prices rose have long been protected by guards. The children of the rich never went to ordinary schools and have long been transported by private buses. At weekends, the rich have always swum, played tennis or golf, and lunched in private clubs. When sick, they have been treated in private clinics. Segregation is a long-established feature of Latin America's cities. While some new gated communities provide their inhabitants with virtually everything they need, most rich families exit the gates at some time of day.

Some have even suggested gated communities help to integrate rather than to divide society. Salcedo and Torres (2004) argue that the juxtaposition of gated communities and poor neighbourhoods creates opportunities for the poor. Improved infrastructure and services bring advantages. In addition, because the inhabitants of the gated community know the poor in their locality, they exclude those people from their generally negative perceptions of poor people in the rest of the city. It is obvious that the gated rich talk to their domestic servants, gardeners and chauffeurs, which was always the main way in which the rich interacted with the poor (Goldstein 2003).

In some cities, of course, the state has sought to improve shelter conditions by building social housing. But this has rarely helped reduce social segregation because the new housing estates have typically been relegated to areas built on the cheapest land, far from the most prosperous neighbourhoods. And the neoliberal capital subsidy approach, despite its virtues, has sometimes made matters worse. Accurate targeting in Chile, for example, has grouped very poor households into homogeneous neighbourhoods. Since these areas are usually located in the least desirable areas of the city and because it is extremely difficult to sell homes in these areas, few families will ever manage to leave. New communities of the poor have been created miles from anywhere. As García Peralta and Hofer (2006, 8) argue: "Putting a socially homogeneous group in a remote and isolated place, and stretching people financially to their limits in a society where a job can easily be lost and an illness can threaten a family's income, produces fear and distrust."

Of course, it is not only the rich who want to cut themselves off. People of every social class seem to be "walling and gating." Poor communities in Bogotá and Santiago sometimes build walls or erect fences around their settlements in an attempt to keep crime and violence at bay. Developers of social housing in Bogotá tell me that they put fences round their apartment complexes because no one will buy the property otherwise.

At the opposite end of the spectrum some would argue that ghettos have developed in most of the region's cities. In Kingston, Jamaica, an area of deprivation dating from the colonial period has evolved into "a zone of poverty and violence ... on a scale unmatched elsewhere in the Commonwealth Caribbean" (Clarke 2006, 31–2). In Kingston, violence between the main political parties helped create the problems of the inner city, whereas, in Rio de Janeiro or Medellín, control of drug trafficking was at the root of violence. Elsewhere, extreme poverty has produced marginalized communities; for example, new shanty towns are being formed on the edge of Bogotá by refugees escaping from the violence of the countryside. The growth of such areas can be remedied only if governments take vigorous action to help the poor and to alleviate urban violence. But it is difficult to argue that Latin American cities suffer from ghettoization, indeed, recent thinking has suggested that ghettos are predominantly a US phenomenon (Gilbert 2012; Marcuse and van Kempen 2000; Wacquant 2012).

Despite the development of gated communities, the research evidence suggests that residential segregation in Latin America has not increased in recent years and is certainly nowhere near as severe as in the United States. "Typically, in Latin American cities the elites represent just a third of the population of high-income areas. Even at the block level, the isolation index of the elites in Santiago did not exceed 40% in 2002. American suburbs, on the contrary, are more socially homogeneous both in racial and income composition, and this homogeneity is more stable through time" (Roberts and Wilson 2009, 133).

Urban Governance

For centuries most Latin American cities were governed as part of a single administrative area. However, as they began to grow outwards, many spilled over into neighbouring administrative areas. The result is that few large cities in the region now have a single government. The current metropolitan area of Mexico City falls into seventy-six different municipal administrations, sixteen in the Federal District, fifty-nine in the state of Mexico and one in the state of Hidalgo. São Paulo is governed by thirty-nine different municipal administrations. Even Bogotá, where the bulk of the population has traditionally lived in the Capital District, some 10 per cent of the population now live in neighbouring areas.

The problem with administrative decentralization in this form is that urban management is now in the hands of many municipalities that are less than competent. They are very different from the authorities in the central administrative areas, which have had decades of experience in urban management and have been able to learn from their past errors. In contrast, the authorities managing the new periphery are much less experienced and are less able to provide basic services or to control the worst excesses of urban development. Indeed, in greater São Paulo some municipalities have sought to attract developers by applying lower environmental and urban standards and have failed to levy the betterment taxes that were contemplated in the widely admired "City Statute" (Bonduki 2011). In Mexico, efforts by the Federal District of Mexico to enforce planning regulations and to protect physically dangerous or ecologically valuable land from urban development encouraged investors to shift to the State of Mexico. There they have been able to find weak politicians who can arrange building permits and do not hinder capital with demands to install essential infrastructure or social services (García Peralta and Hofer 2006).

Conclusion

Latin American cities contain vast swathes of suburbia. Indeed, the vast bulk of Latin American city dwellers are suburbanites. The rapid increase in urban populations during the twentieth century, allied to improved transport and infrastructure provision, guaranteed that all urban growth took the form of suburban development. Of course, Latin American suburbia is highly diverse, ranging from elite neighbourhoods to flimsy self-help settlements. But it is difficult to argue that it has made much difference to social segregation, insofar as Latin American cities have always been among the most unequal places on earth.

Of course, suburban expansion is in many respects unsustainable – urban sprawl is energy intensive, more cars create insufferable traffic jams, air and water pollution is affecting more and more people and too many urban areas are spreading into administrative jurisdictions where the authorities are incapable of properly servicing and organizing the consequent urban sprawl. The failure of most governments to raise enough taxes or to benefit from rising land values by levying betterment charges has held down revenues and prevented them from redistributing resources.

Perhaps suburbanization is a curse, but, since living standards and life expectancy are generally on the rise and most of the problems can be attributed to other causes, it can easily be argued that it is actually producing healthier cities. As in the rest of the world, cities are coping but could be much better organized for the benefit of all.

REFERENCES

Abramo, P., ed. 2009. *Favela e mercado informal*. Porto Alegre: Coleção Habitare.

Abreu, M. de A. 1987. *A evolução urbana do Rio de Janeiro*. Rio de Janeiro: Instituto de Planeamiento Municipal.

Aguilar, A.G., and C. Alvarado. 2005. "La reestructuración del espacio urbano de la Ciudad de México. ¿Hacia la metrópoli multinodal?" In *Procesos metropolitanos grandes ciudades. Dinámicas recientes en México y otros países*, ed. A.G. Aguilar, 265–308. Mexico: Porrúa.

Almarza, S. 1997. "Financiamiento de la vivienda de estratos de ingresos medios y bajos: la experiencia chilena," CEPAL Serie Financiamiento del Desarrollo, 46, CEPAL, Santiago.

Amato, P. 1970. "Elitism and Settlement Patterns in the Latin American City." *Journal of the American Institute of Planners* 36 (2): 96–105. https://doi.org/10.1080/01944367008977290.

Angel, S., J. Parent, D.L. Civco, and A.M. Blei. 2010. "The Persistent Decline in Urban Densities: Global and Historical Evidence of 'Sprawl.'" Working Paper, Lincoln Institute of Land Policy, Cambridge, MA.

Arellano, J.P. 1982. "Políticas de vivienda popular: lecciones de la experiencia chilena." *Coleccion Estudios CIEPLAN* 9: 41–73.

Blanco, A., Gilbert, A.G., and Kim, J.-L. 2016. "Housing Tenure in Latin American Cities: The Role of Household Income." *Habitat International* 51: 1–10.

Bonduki, N.G. 1983. "Habitaçao popular: contribuiçao ao estudo da evoluçao urbana de São Paulo." In *Repensando a habitacao no Brasil*, ed. Lícia do Prado Valladares, 135–68. Rio di Janeiro: Zahar.

Bonduki, N. 2011. "La nueva política nacional de vivienda en Brasil: Desafíos y limitaciones." *Revista de Ingeniería* 35: 88–94.

Caldeira, T. 2000. *City of Walls: Crime, Segregation and Citizenship in São Paulo*. Berkeley: University of California Press.

Clarke, C. 2006. "From Slum to Ghetto: Social Deprivation in Kingston, Jamaica." *International Development Planning Review* 28 (1): 1–34. https://doi.org/10.3828/idpr.28.1.1.

Collier, D. 1976. *Squatters and Oligarchs: Authoritarian Rule and Policy Change in Peru*. Baltimore: Johns Hopkins University Press.

De Mattos, C.A. 1999. "Santiago de Chile, globalización y expansión metropolitana: lo que existía sigue existiendo." *Revista EURE* 25 (76): 29–56.

de Ramón, A. 1985. "Vivienda." In *Santiago de Chile: characteristicas histórico ambientales, 1891–1924*, ed. A. de Ramón and P. Gross, 79–94. London: Monografías de Nueva Historia, Institute of Latin American Studies.

de Soto, H. 2000. *The Mystery of Capital: Why Capitalism Triumphs in The West and Fails Everywhere Else*. New York: Basic Books.

Doebele, W. 1975. "The Private Market and Low-Income Urbanization in Developing Countries: The 'Pirate' Subdivision of Bogotá." Discussion Paper D, 75–11, Harvard University, Department of City and Regional Planning.

Ducci, M.E. 1997. "Chile: el lado obscure de una política de vivienda exitosa." (EURE) *Revista Latinoamericana de Estudios Urbanos-Regionales* 23: 99–115.

Fernandez-Armesto, F. 2013. "Latin America." In *The Oxford Handbook of Cities in World History*, ed. P. Clark, 483–503. Oxford: Oxford University Press.

García-Peralta, B., and A. Hofer. 2006. "Housing for the Working Class on the Periphery of Mexico City: A New Version of Gated Communities." *Social Justice* 33 (3): 129–41.

Gilbert, A.G. 1978. ""Bogotá: Politics, Planning and the Crisis of Lost Opportunities." *Latin American Urban Research* 6: 87–126.

Gilbert, A.G. 1981. "Pirates and Invaders: Land Acquisition in Urban Colombia and Venezuela." *World Development* 9 (7): 657–78. https://doi.org/10.1016/0305-750X(81)90069-3.

Gilbert, A.G. 1998. *The Latin American City*. 2nd ed. London: Latin America Bureau and Monthly Review Press.

Gilbert, A.G. 1999. "A Home Is For Ever? Residential Mobility and Home Ownership in Self-Help Settlements." *Environment & Planning A* 31 (6): 1073–91. https://doi.org/10.1068/a311073.

Gilbert, A.G. 2002a. "Power, Ideology and the Washington Consensus: The Development and Spread of Chilean Housing Policy." *Housing Studies* 17 (2): 305–24. https://doi.org/10.1080/02673030220123243.

Gilbert, A.G. 2002b. "On the Mystery of Capital and the Myths of Hernando de Soto." *International Development Planning Review* 24 (1): 1–19. https://doi.org/10.3828/idpr.24.1.1.

Gilbert, A.G. 2004. "Helping the Poor through Housing Subsidies: Lessons from Chile, Colombia and South Africa." *Habitat International* 28 (1): 13–40. https://doi.org/10.1016/S0197-3975(02)00070-X.

Gilbert, A.G. 2012. "On the Absence of Ghettos in Latin American Cities." In *The Ghetto: Contemporary Global Issues and Controversies*, ed. R. Hutchison and B. Haynes, 191–224. New York: Perseus Books.

Gilbert, A.G. 2016. "The International Experience of Rental Housing." *Habitat International* 54: 173–81.

Gilbert, A.G., and A. Varley. 1991. *Landlord and Tenant: Housing the Poor in Urban Mexico*. London: Routledge. https://doi.org/10.4324/9780203318164.

Giraldo, F. 1994. "La vivienda de interés social: poco subsidio y nada de equidad." *Revista Camacol* 58:14–9.

Giraldo, F. 1997. "Las políticas de vivienda en los noventa." *Desarrollo Urbano en Cifras* 2: 178–229.

Goldstein, D.M. 2003. *Laughter out of Place: Race, Class, Violence, and Sexuality in a Rio Shantytown*. Berkeley: University of California Press.

Held, G. 2000. "Políticas de viviendas de interés social orientadas al mercado: experiencias recientes con subsidios a la demanda en Chile, Costa Rica y Colombia." CEPAL Serie Financiamiento del Desarrollo, 96.

Hinchcliffe, T. 2005. "Review Essay: Elusive Suburbs, Endless Variation." *Journal of Urban History* 31 (6): 899–906. https://doi.org/10.1177/0096144205276993.

Iracheta, A. 2011. ""Experiencias de política habitacional en México," *Revista de Ingeniería, Universidad de los. Andes* 35: 95–9.

Janoschka, M. 2002. "El nuevo modelo de la ciudad latinoamericana: fragmentación y privatización." *EURE. Revista Latinoamericana de Estudios Urbano Regionales* 27 (85): 11–29.

Jauhiainen, J.S. 2013. "Suburbs." In, *The Oxford Handbook of Cities in World History*, ed. P. Clark, 791–808. Oxford: Oxford University Press. https://doi.org/10.1093/oxfordhb/9780199589531.013.0042.

Jha, A.K. 2007. "Low-income Housing in Latin America and the Caribbean," *En Breve*, 101. http://siteresources.worldbank.org/INTLACREGTOPURBDEV/Resources/Jan07LowIncomeHousingEN101.pdf.

Kimm, P. 1993. "Políticas de vivienda, cooperación internacional e integración interamericana: el papel de USAID." In *Anales de la XXXI Conferencia Interamericana para la vivienda*, 47–55. San José, Costa Rica: UNIAPRAVI.

Klak, T. 1992. "Excluding the Poor from Low Income Housing Programs: The Roles of State Agencies and USAID in Jamaica." *Antipode* 24 (2): 87–112. https://doi.org/10.1111/j.1467-8330.1992.tb00431.x.

Klaufus, C. 2010. "The Two ABCs of Aided Self-Help Housing in Ecuador." *Habitat International* 34 (3): 351–8. https://doi.org/10.1016/j.habitatint.2009.11.014.

Kowarick, L. and Ant, C. 1988. "Cem anos de promiscuidade: o cortiço na cidade de São Paulo." In *As lutas sociais e a cidade*, ed. L. Kowarick, 49–74. São Paulo: Passado e presente, Paz e Terra.

Laun, J.I. 1977. "El estado y la vivienda en Colombia: análisis de urbanizaciones del Instituto de Crédito Territorial en Bogotá." In *Vida urbana y urbanismo*, ed. C. Castillo, 295–334. Bogotà: Instituto Colombiano de Cultura.

Marcuse, P., and R. van Kempen. 2000. "Conclusion: a changed spatial order." In *Globalizing cities: a new spatial order?* ed. P. Marcuse and R. van Kempen, 249–75. Oxford: Blackwell. https://doi.org/10.1002/9780470712887.ch12.

Marques, E., and S. Bitar. 2002. "Espaço e grupos sociais na metrópole paulistana." *Novos Estudos* 64: 123–31.

Matos Mar, J. 1968. *Urbanización y barriadas en América del Sur., Gobierno del Estado de Guanajuato, COVEG*. Mexico: Instituto de Estudios Peruanos Mexiquense.

Mayo, S.K. 1999. "Subsidies in Housing." Inter-American Development Bank, Sustainable Development Department Technical Papers Series.

Meyer, R.M., M.D. Grostein, and C. Biderman. 2004. *São Paulo Metropole*. São Paulo: EDUSP and Impren Sãoficial.

Mills, E.S., and J.P. Tan. 1980. "A Comparison of Urban Population Density Functions in Developed and Developing Countries." *Urban Studies* 17 (3): 313–31. https://doi.org/10.1080/00420988020080621.

Monkkonen, P. 2011. "The Housing Transition in Mexico: Expanding Access to Housing Finance." *Urban Affairs Review* 47 (5): 672–95. https://doi.org/10.1177/1078087411400381.

Morris, A. 1978. "Urban Growth Patterns in Latin America with Illustrations from Caracas." *Urban Studies* 15 (3): 299–312. https://doi.org/10.1080/713702382.

Morrison, A. n.d. *The Tramways of Mexico City (Ciudad de México)*. Accessed 14 June 2016. http://www.tramz.com/mx/mc/mc00.html.

Paquette-Vassalli, C. 1998. "Le logement locatif dans les quartiers populaires de Santiago du Chili: les raisons d'un essor limite." Thèse de Doctorat en Urbanisme et Aménagement, École Nationale des Ponts et Chaussées, Marne-la-Vallée.

Pérez-Iñigo González, A. 1999. "El factor institucional en los resultados y desafíos de la política de vivienda de interés social en Chile." CEPAL Serie Financiamiento del Desarrollo, 78.

Puebla, C. 2011. "Las prácticas de los promotores privados para la realización de desarrollos comercial." In *Suelo para infraestructura*, ed. A. Iracheta and E. Soto. Guanajuato, Mexico: Comisión de Vivienda del Estado de Guanajuato.

Ray, T. 1969. *The Politics of the Barrios of Caracas*. Berkeley: University of California Press.

Roberts, B.R., and R.H. Wilson, eds. 2009. *Urban Segregation and Governance in the Americas*. New York: Palgrave Macmillan. https://doi.org/10.1057/9780230620841.

Rojas, E., and M. Greene. 1995. "Reaching the Poor: Lessons from the Chilean Housing Experience." *Environment and Urbanization* 7 (2): 31–49. https://doi.org/10.1177/095624789500700217.

Sa, L. 2007. *Life in the Megalopolis: Mexico, São Paulo*. London: Routledge.

Salcedo, R., and A. Torres. 2004. "Gated Communities in Santiago: Wall or Frontier?" *International Journal of Urban and Regional Research* 28 (1): 27–44. https://doi.org/10.1111/j.0309-1317.2004.00501.x.

Sargent, C.S. 1993. "The Latin American City." In *Latin America and the Caribbean: A Systematic and Regional Survey*, ed. B.W. Blouet and O. Blouet, 172–216. 2nd ed. New York: Wiley.

Scobie, J.R. 1964. *Argentina: A City and a Nation*. Oxford: Oxford University Press.

Scobie, J.R. 1974. *Buenos Aires: Plaza to Suburb, 1870–1910*. Oxford: Oxford University Press.

SDP (Secretariat of District Planning). 2010. *Densidades urbanas: el caso de Bogotá, Bogotá Ciudad de Estadísticas*. Bogota: SDP.

Suárez, M., and J. Delgado. 2009. "Is Mexico City Polycentric? A Trip Attraction Capacity Approach." *Urban Studies* 46 (10): 2187–211. https://doi.org/10.1177/0042098009339429.

UN-HABITAT. 2010. *Urban Trends: Urban Sprawl Now a Global Problem*. Nairobi: UN-HABITAT.

UNDESA. 2011. *World Urbanization Prospects: The 2011 Revision*. New York: UNDESA .

UNECLAC. 2007. *Statistical Yearbook for Latin America and the Caribbean*. Santiago: UNECLAC.

UNECLAC. 2012. *Social Panorama for Latin America and the Caribbean*. Santiago: UNECLAC .

Vaughan, L., S. Griffiths, M. Haklay, and C.E. Jones. 2009. "Do the Suburbs Exist? Discovering Complexity and Specificity in Suburban Built Form." *Transactions of the Institute of British Geographers* 34 (4): 475–88. https://doi.org/10.1111/j.1475-5661.2009.00358.x.

Wacquant, L. 2012. "A Janus-faced Institution of Ethnoracial Closure: A Sociological Specification of the Ghetto." In *The Ghetto: Contemporary Global Issues and Controversies*, ed. R. Hutchison and B. Haynes, 1–32. New York: Perseus Books.

Wikipedia. 2013. *Rascacielos en Bogotá*. Accessed 14 June 2016. https://es.wikipedia.org/wiki/Anexo:Rascacielos_en_Bogot%C3%A1

Xavier, H.N., and F. Magalhães. 2003. "The Case of Rio de Janeiro, UNDERSTANDING SLUMS: Case Studies for the Global Report on Human Settlements 2003." Accessed 14 June 2016. http://www.ucl.ac.uk/dpu-projects/Global_Report/pdfs/Rio.pdf

chapter ten

Urban Governance, Land Use, and Housing Affordability: A Transatlantic Comparison

FRANÇOISE JARRIGE, EMMANUEL NÉGRIER, AND MARC SMYRL

The question of land use in suburban settings is a fundamentally political one. As such, the challenge of the collective control of this resource is one that must be faced by all political authorities. Answers to it, however, tend to be locally differentiated, as different systems of politics and institutions put forward a variety of formulas for managing the diverse and often conflicting interests and goals involved. In suburban settings in both Europe and North America, moreover, the most politically expedient answers have tended to bring forth additional questions, introducing broader issues from optimum land use to the trade-off between market efficiency and social justice. The political institutions called on to address these issues vary considerably across national settings, as do the ideologies that underlie them. In this study we explicitly choose examples from two states, France and the United States, generally held to be quite far apart in many important respects. French centralism is contrasted with the complex mix of local and federal authority in the United States; American emphasis on the positive moral value of markets is expected to bring results markedly different from the public commitment of France to the use of state authority in pursuit of a more egalitarian social vision.

In this context land-use policy makes an interesting institutional comparison for its own sake, but also with respect to the dynamics of its implementation and its interaction with other policy goals, such as the protection of agricultural land from urban sprawl, the respective roles of state and markets, and the challenge of providing affordable housing. In this chapter we have sought to consider the relationship among these elements, looking particularly at the interaction between institutions and policies governing the use of suburban land and the question of housing affordability. To this end, we focus on the impact of three elements: existing patterns of settlement; public finance in the broadest sense, including both sources of income to relevant authorities and the impact of their spending choices; and the geographical organization of government (Harris, Lehrer, and Bloch 2012). The two urban areas chosen for comparison – Montpellier in France and Aurora, Colorado, in the United States – have in common a recent history of rapid and sustained suburban growth and a substantial population of households with modest incomes. They differ significantly, however, with respect to both the structures and the norms of governance. Thus, taken together, this is a comparison that allows us to consider the relative impact of formal institutions, on the one hand, and economic and geographical dynamics, on the other, on the question of housing affordability in a rapidly growing suburban setting. In the first section of this chapter, we sketch out these general similarities and differences then turn to a discussion of their impact on housing policies.

Aurora and Montpellier: Patterns of Existing Settlement

It is a truism that no public policy begins with a clean slate. With respect to housing and land-use issues, patterns of settlement as they have developed over time provide not only a mandatory starting point but, in many cases, limits to possible change. The two cases chosen here for study, as we will see, share a history of rapid urban growth in the late twentieth century that provides a common setting for policy in the twenty-first century.

Owing to the attractiveness of its Mediterranean coastal location, the city region of Montpellier in Occitanie in the south of France is a dynamic area whose population growth is fuelled by positive net migration. Urban sprawl has caused significant changes in the exurban landscape during recent decades, individual housing tracts spreading out from the core of all villages. New building has taken place in areas where vineyards previously were the characteristic feature. The local economy now largely depends on the service sector, but regional levels of unemployment remain persistently higher than the French average. The combined effects of the land release triggered by the wine (over) production crisis as well as population growth has led to an accelerated consumption of rural land in the exurban fringes of Montpellier (Table 10.1 and Figure 10.1).

Table 10.1. Population growth and urban sprawl in Montpellier city region, France, 1960–2004

1960	1980	2000	2004
145,000 inhabitants	280,000 inhabitants	375,000 inhabitants	395,000 inhabitants
Urban area: 1,000 ha	Urban area: 4,000 ha	Urban area: 9,000 ha	Urban area: 10,000 ha
Urban density: 145 inhabitants/ha	Urban density: 70 inhabitants /ha	Urban density: 41 inhabitants/ha	Urban density: 39.5 inhabitants/ha

Source: Montpellier Agglomération (2012)

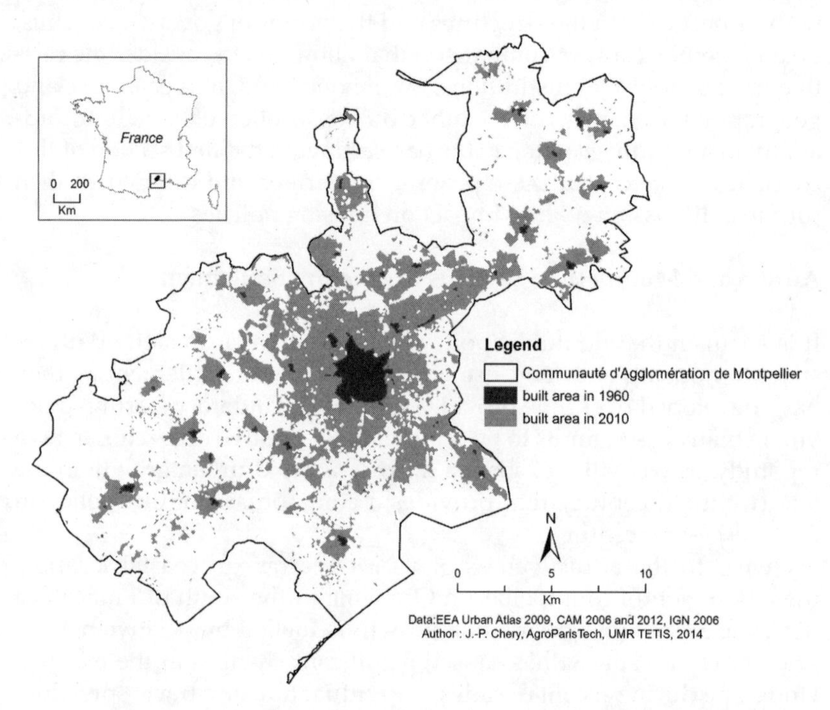

Figure 10.1 Urban sprawl in Montpellier city region, France, 1960–2010

In recent decades the city region of Montpellier experienced one of the highest demographic growth rates in France, but this growth was barely managed at the municipality level, and there was no planning scheme on the scale of the functional urban area (*aire urbaine de Montpellier, 93 municipalities*; INSEE 2010). In consequence, the driving forces of the region's growth led to uncoordinated development with an increasing per capita rate of spatial consumption until the creation of a new inter-municipal authority, the Montpellier Agglomération, in 2001. The city region's authority was further expanded when it became Montpellier Méditerranée Métropole in 2015. The principal instrument of this new metropolitan authority is the *Schéma de Cohérence Territoriale* (SCoT), a regional planning document produced in 2006 and updated in 2017. While not legally binding participating municipalities, this regional planning document has been seen as a first step in harmonizing land-use policies among the municipalities.

The Denver area, of which the city of Aurora is a part, is in many ways typical of the western United States in that its pattern of urban growth is characterized by suburban and exurban sprawl emanating from a small number of original urban centres whose existence dates to the mid-nineteenth or early twentieth century. Also typical is the prevalence of population centres altogether independent of any historical urban centre. Among the most important of these is the city of Aurora, focus of the case study below, which covers portions of three counties and, with a 2010 census population of 230,000, is second only to the city of Denver in population size.

While the city of Denver is largely constrained in its growth by its central position in its urban area, Aurora is situated squarely on the exurban fringe. An aggressive policy of annexation has led to the inclusion within the city limits of large expanses of agricultural land. The extension of urbanization (urban sprawl) has been a constant for Aurora and its surrounding area (at least) since the 1970s. Figures 10.2a and 10.2b below illustrate this trend.

Structures of Governance

As noted in the introduction and further detailed below, the models of land policy represented by Montpellier and Denver are starkly different at the level of formal institutions. It does not follow, however, that they cannot usefully be compared. Comparability does not require, or even suggest, similarity in every aspect. On the contrary, it invites us to look for and to explain differences. A first obvious difference, as noted above, is that urban boundaries in the United States are much less fixed than is the case in France, where all national territory is divided into townships. As Aurora did in the twentieth century, American cities can expand their boundaries aggressively through annexation of hitherto unincorporated land. A second obvious difference is the diversity of governance regimes available to local authorities in the US case. Whereas all French *communes* of a given size are governed, in principle

Figure 10.2a Arapahoe County, USA: population density, 1990.
Source: Arapahoe County Master Plan

Figure 10.2b Arapahoe County, USA: population density, 2010.
Source: Arapahoe County Master Plan

at least, by identical rules, US conditions vary greatly. As a "home rule" city in the State of Colorado, Aurora largely escapes the direct supervision of intermediate echelons of local government such as the counties in which it is situated.

However, as we will see, this institutional difference can mask deeper similarities. Aurora may be politically quite autonomous from its county and state governments, but in certain policy areas – and affordable housing is such a one – it is deeply dependent on US federal authorities both for resources and for regulatory structure. The townships of the Montpellier region, for their part, exercise a surprising degree of de facto autonomy despite a seemingly much more rigid institutional structure. An initial comparison of the formal planning process, and its limits, in the two cases provides a useful starting point.

Land-use policy in France reveals a highly voluntaristic system of public intervention in which authorities hold a wide array of regulatory resources and instruments. At the same time, however, the system is extremely fragmented and provides far more influence than might first appear to private developers and builders. Public intervention in this area is made possible by several instruments. The first is the *Plan Local d'Urbanisme* (PLU) produced by the majority of municipalities. Designed by local authorities in accordance with national law as well as with the SCoT – as previously seen – the PLU draws up detailed zoning guidelines defining allowable uses for each parcel of land. About a third of municipalities, generally the least populated, do not have such a plan, in which case the zoning authority reverts to the national government, which in any event maintains the power to impose infrastructure projects on municipalities. Additional instruments of public authority include the power to acquire property by right of eminent domain for projects deemed to be in the public interest. The construction of public housing can fall into this category. A final relevant instrument is the right of pre-emption of private sales, held by municipalities and *départements*, but also by quasi-public entities such as the *Société d'Aménagement Foncier et d'Etablissement Rural* (SAFER), which has the authority not only to pre-empt a sale, but also to lower the price if deemed necessary to slow speculation on agricultural land.

For all of the seeming authority provided by these instruments, however, the extreme fragmentation of public authority and the considerable variation in the concrete planning capacity of local units can result in land-use policies being limited to a set of formal rules and the power to issue building permits. This restriction gives considerable strategic advantages to large private sector actors, who often have a more unified economic and commercial strategy, especially in areas of high demographic pressure. In such cases, builders and developers can accumulate a large stock of undeveloped land by taking advantage of the economic difficulties of agriculture, which sometimes leads to a situation in which private actors wield dominant influence over zoning and land-use policies. This state of affairs, as we shall see, is not so different from that of our US case.

Arapahoe County produces a "Master Plan" for land use that, like the SCoT in Montpellier, has indicative rather than mandatory force. As in the French case, however, development projects in accord with the Master Plan are likelier to meet with planning approval. Just as developers are required to set aside land for environmental reasons, in principle they could be required to build a mix of housing, including some at moderate cost. Nevertheless, the general practice in the state of Colorado, as exemplified by Arapahoe County is not to enforce this requirement. At most, the planning staff may be predisposed to approve development projects likely to sell or rent at lower prices.

Municipalities such as Aurora have a broader set of tools at their disposal; their home-rule status, granted by the state legislature, largely removes them from the authority of their home county. Even so, the question then becomes about how these tools are used. In the state of Colorado municipal governments may choose to offer financial incentives on a case-by-case basis to encourage favoured types of development. The city of Aurora regularly uses these incentives to encourage economic growth, typically by encouraging the establishment of potential employers. In principle, similar inducements could be used to promote the construction of affordable housing, but there is little evidence that any such effort is being made.

Access to Housing: Comparing the Two Modes of Land Governance

The Montpellier city region's growth-management strategy centres on controlling sprawl while at the same time meeting the need for affordable housing. Its objective, according to the local housing plan (PLH) is the production of 5,000 units per year, of which a quarter would be public housing. As a result of rapid population growth in the exurban belt around Montpellier, a growing number of small municipalities are required, in principle, to provide social housing as their population increases beyond the 3,500 threshold. In this, the Montpellier area mirrors the broader French demographic context. Between 1968 and 2008 the exurban zone witnessed average annual population growth of more than 1.1 per cent compared with effectively zero growth for urban cores. This growth has gone hand in hand with the expansion of owner-inhabited single-family detached houses. Since 1976 the production of single-family dwellings has surpassed that of multi-family units (Madoré 2012). Eighty per cent of such dwellings are individually owned.

This hegemony of single-family housing consumes a great deal of space, but it also has social and political consequences. On the social level the dominant representation is that of a strong desire for social homogeneity (Charmes 2011). In this context development of "social housing" is extremely difficult, even when it is legally required. Of the thirty-one municipalities in the city region, only twelve explicitly include the goal of mixed-income housing in their PLU. To the extent that this mindset is evolving at all, it is because the rise in property values makes it increasingly difficult for the local young adults to find affordable housing nearby. Support for "social housing" thus actually consists in these areas of attempts by the existing population to maintain itself, its offspring, or people as much like themselves as possible in a locally dominant position. This is, of course, a general tendency, to which local exceptions can be found. What is increasingly clear, nevertheless, is that the spread of the individual owner-resident model and its adoption by households of modest means have a direct effect on the expansion of demand for urbanized land.

Historically, the production of social housing was concentrated in the city of Montpellier itself. For this reason Montpellier has a concentration of lower-income population – or at least that portion of it that is fortunate enough to obtain access to social housing. For other low-income households, despite their eligibility in principle for social housing, there was no option other than the private market. As a result, the near-universal pattern is one of concentric circles, where the price of housing declined with increased distance from the urban core. Lower-income households seeking detached housing were thus forced to seek housing farther and farther from the city centre, increasing transportation expenses, sometimes to the point of endangering their financial stability.

This case is not unique to Montpellier. It is found in the Paris region (Aragau, Berger, and Rougé 2012) as well as elsewhere in Occitanie. Even so, the phenomenon is particularly marked in this region, as it is both one of the fastest-growing and one of the poorest in France. The impact of the situation on low-income households is both geographical, since the lack of social housing forces them ever farther from the urban centre, and financial, as housing costs account for ever more of household budgets. Between 2005 and 2010 the cost of housing increased by 17 per cent overall, but this increase was 26 per cent for homebuyers, while it was limited to 9 per cent for renters of social housing.

The evolution of the French housing model therefore is ambivalent. On the one hand, it is distinguished by the production of housing by public authorities, resulting in a genuine impact on the cost of housing, and thus on access to housing, for low-income households. This production, nevertheless, is clearly insufficient and allows the parallel development of a market that public authorities have difficulty in regulating. Rising housing prices that regularly surpass overall inflation rates illustrate this problem: those low-income households without access to social housing must search for cost-minimizing strategies such as seeking lower-cost housing far from the centre. Such places often prove extremely burdensome for the households that pursue them, in addition to being an important factor contributing to urban sprawl and the loss of agricultural land. This phenomenon, finally, has a clear political impact, as it engenders multiple frustrations for new as well as existing inhabitants of exurban areas, including the shift to the extreme right as the fastest growing electoral support (Négrier 2012).

In this ambiguous context, the Montpellier city region might seem like a "good student" in recent years, since the implementation of the SCoT. Between 2007 and 2012 an average of 5,040 housing units were added per year, of which 21 per cent was public housing. Even so, despite efforts at urban reinvestment and increasing density, no significant decrease in the rate of urban expansion was noted (Montpellier Agglomération 2012). These results shine a light on the diversity of municipal strategies within the Agglomération. Aggregate numbers fail to show the extent to which urbanization has increasingly been shifted to outlying municipalities that come under the influence of the city's economic growth, but whose housing policies are not subject to regulatory pressures of instruments such as the SCoT or the PLH.

In the exurban universe a logic of social separatism and a spatial hierarchy is in the process of establishing itself in a pattern of concentric circles around desirable urban cores. Weak respect for formal rules, or the hijacking of policy instruments in the name of social and familial reproduction in smaller and medium-sized municipalities, provides an image quite different from the official one of "France, home of land-use regulation." When we turn our gaze across the Atlantic we note obvious differences but also, as will be discussed below, surprising areas of convergence.

At the outset of this chapter we stated that the determination of land use in general was an intensely political question, both in the United States and in France; it proves to be even more strongly the case for provision of affordable housing. Various jurisdictions will adopt a very different official position on the issue. Housing affordability is all the more important to the American policy debate in the wake of the mortgage crisis of 2008. An intense public debate on housing affordability, however, does not necessarily translate into a direct policy role for public authority. Unlike retirement or income support and healthcare for low-income persons, housing has never been considered an entitlement by the US federal government. Meanwhile, questions of land use, as we have seen, remain almost purely local. The result is a variety of uncoordinated policy initiatives on the part of federal and local authorities that take aim at one or another of these issues often with little consideration for its impact on the others.

"Affordability" is necessarily a subjective notion. By some measures, housing in the Denver-Aurora area is relatively affordable. Data from the US National Association of Realtors (2013) indicate that median family income in the metropolitan area stands at 163 per cent of the level considered adequate to qualify for a mortgage on a median home. The comparable 2011 figure for the western region of the United States was 118 per cent, and for the United States as a whole 186 per cent. However, behind averages of this kind lies a more complex reality. Expenses related to housing make up a disproportionate part of the budget of low-income households. Data collected by the US Bureau of Labor Statistics (2013) suggest that, for the lowest-income quintile of the US population in 2011, housing, which includes rent or mortgage interest as well as utilities and other housing-related spending, stood at nearly 40 per cent. This very different measure of housing affordability clearly shows the inverse relation between income and the cost burden of housing, the proportion of household budgets going to housing decreasing to 35 per cent in the middle quintile and 31 per cent for the wealthiest fifth of the American population. If we take housing expenses as a proportion of income rather than of expenditure, the difference is more striking. As a proportion of reported taxable income, housing expenses range from nearly 90 per cent for the poorest quintile to 18 per cent for the wealthiest. Generalizing from such data, Quigley and Raphael (2004, 192) conclude that over the four decades leading to the 2000s, the United States was characterized by "pronounced increases in the typical rent burden for poor and near-poor households." However "affordable" housing may be for the middle-class, homeowning family, the cost of shelter can be a heavy burden indeed for those at the bottom of the income distribution. Housing policy seeks to address this problem, but its means for doing so are limited.

Policy initiatives aimed at improving access to affordable housing have historically taken one of three possible approaches. Best known to middle-class Americans are initiatives intended to make the purchase of a home financially less burdensome. In some cases local planning agencies can offer inducements to the production of modestly priced housing. Most direct, finally, are the housing assistance policies of the federal Department of Housing and Urban Development (HUD). In the conditions obtaining in Aurora all three initiatives contribute significantly to sprawl.

The indicative planning tools available at both the county and the municipal levels have typically been used to promote economic growth, focusing particularly on employment. In a land-rich setting such as Aurora, access to relatively low-cost formerly rural real estate is put forward as a competitive advantage for businesses relative to the more congested central metropolitan zone. At the county level planners may be predisposed to approve residential development projects likely to sell or rent at lower prices, but, as already noted in the case of Montpellier, this attitude almost always implies moving ever farther from the urban centre. In both cases, moreover, the financial incentive is clear: development is directly linked to increased tax revenue and, other things being equal, is deemed desirable. Whereas sprawl is often seen as an unintended consequence of urban policy (Blais 2010), we should be prepared to consider the possibility that, in this case, it is a direct and intentional result. This starting point, moreover, has consequences on the impact of policies that initially were meant to address an altogether different problem.

The most common and direct form of housing aid to low-income households in the United States is through the Housing Choice Vouchers program funded through the budget of the federal Department of Housing and Urban Development and administered by local public housing authorities. Universally known to beneficiaries as well as to administrators as Section 8, the vouchers allow families with income below 50 per cent of the area's median income (AMI) to rent housing on the private market. A proportion are reserved for families considered "very low income": those whose annual revenues are below 30 per cent of the AMI.

It is important to note from the outset that Section 8 is not an entitlement; it is allocated on the basis of waiting lists established and administered by local public housing authorities. Budgetary reforms at the federal level in the 2000s have strengthened this feature of the program, effectively capping overall funding available at the national level without regard to changes in need. Unlike entitlements, moreover, funding for Section 8 is subject to annual review by Congress and is at risk when Congress and the president fail to agree on budgets, as has happened many times since 2010. For the nation as a whole, current funding provides assistance to approximately 2 million households.

For the Public Housing Authority of Aurora this translates into an allotment of approximately 1,300 contracts. An additional number of households, approximately 400, live in the city of Aurora but receive Section 8 vouchers from another PHA. The principal limitation of the Section 8 program, accordingly, is budgetary. The Aurora Housing Authority's stock of contracts is defined historically, based on decisions going back as far as the 1970s. There is no regular reassessment of actual need. The result is a supply of vouchers substantially smaller than the potential demand for them.

This reliance on the private market as a source of supply, so evident in Aurora, is representative of a national trend since the 1970s. A number of reasons can be put forward for the change of approach, chief among which at the national level was the perceived failure of certain inner-city high-rise "projects" in the 1960s and 1970s, which developed a reputation for creating concentrations of extreme poverty and associated social problems. While these conclusions have been a subject of debate among researchers (Galster and Zobel 1998; Crump 2003), their impact on practical policy-making is undeniable. Even in places such as Aurora, where "mega-projects" and their associated dysfunction never existed, the feeling among administrators, confirmed in interviews for this project, is that voucher-based programs are socially superior in that they encourage a higher degree of social and economic diversity in neighbourhoods, avoiding concentrations of assisted households. When brought together with the other elements we have seen, a clear pattern emerges. The type of mixed-income neighbourhood considered ideal can exist only by encouraging landlords in neighbourhoods initially characterized by mid-level housing to accept Section 8 contracts. In a city such as Aurora these areas are, more or less by definition, widely dispersed. Acting to prevent concentrations of poverty may be a socially laudable goal, but under the conditions obtaining here one clear consequence is that sprawl begets more sprawl.

Comparative Discussion

The result of our comparison allows us to identify points both of convergence and of abiding differences. The analytic lessons we draw relate at once to the notions of culture, of implementation, and of change.

Looking first at the points of divergence between our two cases, we note that they are evident at the most fundamental and longest-term dimensions of public policy: their constitutive level. Here, we are dealing with the fundamental norms and values that dominate and define the fields of real estate and land use. They include the overall legitimacy of public actors in the system and the limits (if any) imposed on the private market. Looking at the two systems, one can clearly contrast the legitimacy of public institutions at all territorial levels in France with respect to the production of social housing, with the American assignment of much greater responsibility to private developers, intervening only to support demand. The assumed legitimacy of the market in the American case is mirrored by the generalized assumption of state responsibility in France. It does not follow, however, that there is no low-cost housing in the United States – or that it is provided without question in France. Rather, in both cases it means that it is supplied according to the rules and dynamics of markets and politics.

In the same perspective the notion that public authority has a responsibility to regulate the market, which is translated in France by the tools put at the disposal of the agents both of the national government and of local jurisdictions (planning schemes, projections, land reserves, etc.), seems incongruous to American eyes. We make explicit reference to national scales of comparison here, because the levels of expression of these values are indeed those of national states, with their respective histories and policies.

Nevertheless, the chief difference seems to be one of intent rather than of results. While both cases are characterized by fragmentation in the system of public action, this is seen as a problem in France, where we can see a permanent – albeit largely futile – search for rationalization, simplification, and change of scale. In the American case, on the other hand, the fragmentation of operators, authorities, and services seems to be a basic principle of land-use dynamics. Likewise, with respect to patterns of settlement, sprawl is generally seen as problematic in France, while decision-makers in US cities such as Aurora seem inclined to believe – rightly or wrongly (Lee 2011; Phelps 2012) – that extensive growth, that is, sprawl, is both economically and socially desirable. We must be wary of excessive generalization; our observations of Aurora cannot be seen as universally representative of an undifferentiated American reality. What they do clearly show us is a case in which the developmental and social "faces" of urban policy (Kantor 2013), far from being antithetical, are closely linked. It is not an exaggeration to say that, in Aurora, economic development is social policy. Thus, we seem to be in the presence of two radically different contexts with respect to values and norms and to actors, but these contexts are situated at a high level of abstraction. If we look more closely at implementation outcomes, our observations are much more nuanced. Indeed, we can suggest important elements of convergence.

The first, ironically enough, is with respect to differences. Indeed, in each of the cases studied we can observe important contrasts in implementation of the various policy tools among urban regions or even among local jurisdictions within them – an observation perhaps more surprising in the French case. It is not unexpected to find important differences in policy choices between the city of Aurora and other portions of the greater Denver-Aurora metropolitan area, but we observe equally significant differences between Montpellier and the outlying areas of Occitanie, or indeed between this and other French regions. With respect to implementation, the coherence of national models is thus partial at best, and this incoherence is a first point of convergence between the national cases. What is striking in both cases is the underuse of the available instruments of metropolitan governance; regional planning tools, as we have seen, exist but are difficult to enforce. Local jurisdictions, even very small ones, typically find themselves in a binary relationship with national laws and authorities on this question. Whatever the accomplishments of "metropolitan regionalism" may be in areas such as transportation or culture (Brenner 2002), they do not extend to any significant extent to this policy area.

Looking at the institutional landscape in terms of a broader view of "suburban governance," focusing on the "agents, methods, and institutions through which development is managed," (Ekers, Hamel, and Kiel 2012) brings the two cases even closer, revealing in both France and the United States overlapping and competing fields of activity. In France private developers coexist with public and semi-public ones (the *sociétés d'économie mixte* as well as local public authorities), while US local housing authorities are entirely dependent on federal funding; in both countries different levels of action may have redundant or overlapping authority and responsibility.

A second element of convergence is tied to certain similarities in the way in which the instruments are put in place on the ground. This is a convergence in the strict sense: each of the two systems, while formally opposed, has moved towards the other or, perhaps more to the point, the French reality has come to resemble long-standing policy in the United States. In the case of Montpellier we have seen that townships with more than 5,000 inhabitants have the obligation to devote 20 per cent of their housing units to social housing. In practice, however, this rule is not respected; it is openly contested by certain elected officials and evaded by others. We can thereby deduce a tendency to rely on the market, and very possibly towards social segregation, which would be difficult to see if we limited our attention to official texts or to the formal instruments of public action.

A third point of convergence is found with respect to the actors themselves. One illustration suffices – a comparison between developers in Colorado and *lotisseurs-aménageurs* in Montpellier. These two examples present strong similarities: the same local structure of capital (based on real property) and the same need for political skills, whether an ability to negotiate the local political power structure and social networks or the importance of negotiation. These actors also face similar temporal constraints related to the necessity to tie up capital, the length of projects, and the need to allow for the uncertainty of negotiations.

These elements of convergence, as well as of difference, suggest the necessity of a closer look at the social impact of the two models of policy for households. The piecemeal nature of French social housing policy leads to a split in the situation of low-income households between those who succeed in getting access to social housing and others who do not. While the first continue to represent a reality totally different from the American situation, what can we say of the latter? Dependent on a largely unregulated market driven by a speculative logic, they live in a situation that, in the end, is quite comparable to low-income American households. A comparison of the weight of housing costs in household budgets (41 per cent for the lowest quintile in the United States; 34 per cent in France) shows a difference of degree, but not of nature. In this field, as well, our study allows us to show that the two models, while they are radically different with respect to their guiding principles, seem much more convergent when we look at actors in an interaction, the impact of rules, and the effects in terms of household budgets. Convergence is also evident through the impact on land use – in practice, if not in principle. Both of the cases we observed are, in effect, using sprawl either as a substitute for public provision of affordable housing or, in the American case, as an acknowledged element of such a policy. The relationship between a first group of questions dealing strictly with urban housing policy (urban housing affordability, tools of public intervention in the urban housing market, and strategies of avoidance on the part of civic authorities) and a second group of questions that looks squarely to issues of suburban land use may seem distant at first glance, but our observations suggest that it is, in fact, quite direct.

The lessons we can draw from this comparison are of three orders. In the first place, the formal and informal aspects of land-use policies are clearly interdependent. The informal level is not the sign of a particular culture, or of an absence of capacity for action. Indeed, it can be precisely the reverse: strategies to get around rules need those very rules to operate. In our two cases, where capacity for action, forecasting, and planning are evident, informal action is linked to the distance we perceive between the overall stated objectives of policy, the instruments they create, and their implementation on the ground. The deliberate misuse of regulation, the tendency to exploit ambiguities in levels of authority, and the importance of negotiating skills are examples of such informal behaviour.

Further, despite the convergence that we observe, "land use cultures" are a reality, and not only at the global level. These cultures reflect a system of beliefs that is sufficiently rooted in function without always being consciously expressed, and that can be all the more effective when they dispense with objectivity.

Finally, it is useful to reflect on the meaning of such a comparison. In addition to what we have said so far, it is worth observing this comparison in a dynamic sense. What has always distinguished the two systems is the constraint that American local governments are heavily dependent on locally raised taxes, while the French model gives the state stronger tools of fiscal redistribution across its territory that have allowed local governments to escape from an overly strong dependence on local resources. "Urban growth machines" (Molotch 1993; Jonas and Wilson 1999) or "urban regimes" (Stone 1993, 2005; Stoker and Mossberger 1994) have always been hard to translate into French terms. Even so, in a period of decline in national financing in France, it seems evident that the tendency to instrumentalize local fiscal or land resources will be ever stronger. At the same time, the comparison between the American and the French systems will cease being an exercise in intercultural dichotomy to engender studies of variable concrete convergence.

REFERENCES

Aragau, C., Berger, M., and Rougé. 2012. "Du périurbain aux périurbains: diversification sociale et générationnelle dans l'ouest francilien." *Pouvoirs Locaux* 94: 58–64.

Blais, P. 2010. *Perverse Cities: Hidden Subsidies, Wonky Politics, and Urban Sprawl*. Vancouver: UBC Press.

Brenner, N. 2002. "Decoding the Newest 'Metropolitan Regionalism' in the USA: A Critical Overview." *Cities (London)* 19 (1): 3–21. https://doi.org/10.1016/S0264-2751(01)00042-7.

Bureau of Labor Statistics. 2013. *Consumer Expenditures in 2011 (BLS Report 1042)*. https://www.bls.gov.

Charmes, E. 2011. *La ville émiettée. Essai sur la clubbisation de la vie urbaine*. Paris: PUF.

Crump, J. 2003. "The End of Public Housing as We Know It: Public Housing Policy, Labor Regulation, and the US City." *International Journal of Urban and Regional Research* 27 (1): 179–87. https://doi.org/10.1111/1468-2427.00438.

Ekers, M., P. Hamel, and R. Keil. 2012. "Governing Suburbia: Modalities and Mechanisms of Suburban Governance." *Regional Studies* 46 (3): 405–22. https://doi.org/10.1080/00343404.2012.658036.

Galster, G., and A. Zobel. 1998. "Will Dispersed Housing Programmes Reduce Social Problems in the US?" *Housing Studies* 13 (5): 605–22. https://doi.org/10.1080/02673039883128.

Harris, R., U. Lehrer, and R. Bloch. 2012. "The Suburban Land Question." Paper presented at the Global Suburbanisms Project Workshop on Land, Montpellier, France, 21–23 October 2012.

INSEE. 2010. *Les chiffres-clés de Montpellier Agglomération*. Paris: Institut national de la statistique et des études économiques.

Jonas, A.E. Wilson, D., eds. 1999. *The Urban Growth Machine: Critical Perspectives, Two Decades Later*. Albany, NY: SUNY Press.

Kantor, P. 2013. "The Two Faces of American Urban Policy." *Urban Affairs Review* 49 (6): 821–50. https://doi.org/10.1177/1078087413490396.

Lee, S. 2011. "Metropolitan Growth Patterns and Socio-Economic Disparity in Six US Metropolitan Areas: 19970–2000." *International Journal of Urban and Regional Research* 35 (5): 988–1011.

Madoré, F. 2012. "Habiter la France périurbaine." *Pouvoirs Locaux* 94: 52–7.

Molotch, H. 1993. "The Political Economy of Growth Machines." *Journal of Urban Affairs* 15 (1): 29–53. https://doi.org/10.1111/j.1467-9906.1993.tb00301.x.

Montpellier Agglomération. 2012. *SCoT, bilan d'étape. 6 années d'action pour un développement plus durable de notre territoire*. Accessed 16 June 2014. http://www.montpellier3m.fr/sites/default/files/downloads/files/scot-bilan-2004-2010-07-12.pdf.

National Association of Realtors. 2013. *Housing Affordability Index*. Accessed 16 June 2016. www.realtor.org/topics/housing-affordability-index.

Négrier, E. 2012. "Le Pen et Le Peuple. Géopolitique du vote Front National en Languedoc-Roussillon." *Pôle Sud* 37: 153–66.

Phelps, N. 2012. "The Growth Machine Stops? Urban Politics and the Making and Remaking of an Edge City." *Urban Affairs Review* 48 (5): 670–700. https://doi.org/10.1177/1078087412440275.

Quigley, J., and S. Raphael. 2004. "Is Housing Unaffordable? Why Isn't It More Affordable?" *Journal of Economic Perspectives* 18 (1): 191–214. https://doi.org/10.1257/089533004773563494.

Stoker, G., and K. Mossberger. 1994. "Urban Regime Theory in Comparative Perspective." *Environment and Planning: Government and Policy* 12 (2): 195–212. https://doi.org/10.1068/c120195.

Stone, C.N. 1993. "Urban Regimes and The Capacity to Govern: A Political Economy Approach." *Journal of Urban Affairs* 15 (1): 1–28. https://doi.org/10.1111/j.1467-9906.1993.tb00300.x.

Stone, C.N. 2005. "Looking Back to Look Forward: Reflections on Urban Regime Analysis." *Urban Affairs Review* 40 (3): 309–41. https://doi.org/10.1177/1078087404270646.

chapter eleven

An Effective Public Partnership for Suburban Land Development: Fleurhof, Johannesburg

MARGOT RUBIN AND RICHARD HARRIS

People often speak about suburban development as if the process is inevitable, a tidal sprawl of people, buildings, and infrastructure. But of course, for suburbanization to happen, land has to be sold, subdivided, and developed. There is nothing inevitable about this, and certainly not the way in which it happens. Typically, land is developed either by public or, more commonly, through private agency, but in recent years hybrid Public Private Partnerships (PPPs) have become increasingly common. They have come in for a good deal of criticism, notably on the grounds that in effect they provide a state guarantee for private profits. But they can work well. The South African case study reported here indicates that, with strong political leadership and state guidelines, development partnerships can be a valuable policy and planning tool.

PPPs have been the object of intense debate. A substantial literature has been developed that advocates their advantages, allowing the different sectors to play to their strengths (Jones and Pisa 2000). They can ensure the efficiencies of the market are balanced by state intervention and public participation, which enables the protection and inclusion of marginalized communities (Payne 1999). At the same time, critics maintain that in general very few PPPs have achieved outcomes at any scale or reached poorer groups and communities (United Nations 1993). Some researchers have taken a more balanced view and looked at the benefits and negative consequences of these instruments, discussing their "pitfalls and promises" (Choe 2002) along with their "hope and expectations" (Jones and Pisa 2000). Despite the proliferation of these studies, very few researchers have investigated how PPPs are operationalized and the nature of the partnership at their centre. The following case begins to speak to this gap and offers a conceptual framework that would begin to shine a light into the black box on contemporary writing of PPPs in South Africa.

This chapter narrates the story of the origin and construction of the Fleurhof Integrated Housing Development (hereafter Fleurhof) in Johannesburg, South Africa's largest metropolitan area. Using primary material drawn from a set of interviews (see the appendix for a complete list) conducted between mid-2013 and 2014 with key people from local government and the private sector who were involved in the inception, construction, and development of the project. Respondents were chosen on the basis of direct involvement in some aspect of the project and represented key organizations and institutions that had an interest in the project and its outcomes. We also used secondary-source documents, including media reports, project reports, and local government policy documents. By synthesizing the material we are able to describe the nature of the interactions and intersections between these actors in the development and construction of this new settlement. To aid in the analysis we have constructed a conceptual framework to define the phases of land assembly in South Africa and sketch out some of the dimensions of PPPs. We conclude by discussing some of the advantages such instruments offer to all of the concerned parties.

Fleurhof is emblematic of current plans for the development of suburbs in Johannesburg and other South African cities. Fleurhof is located on an old buffer strip within what has been termed Johannesburg's mining belt, a section of land that has largely been sterilized and seen little to no development for the past century. However, the area has recently been identified as a key site for suburban development and is so targeted by the City of Johannesburg. Furthermore, the site is intended to integrate the geography of a city still divided by a legacy of apartheid spatial policies. The project thus responds both to the national call for infill projects that literally fill in the spaces that have been left over by these old policies and to local calls for housing. The Fleurhof project is also part of a new generation of housing developments, seen as a megaproject of mixed housing and mixed land development, and to which the state has entrusted the resolving of the 1.5-million-unit backlog. In late 2014 the minister of human settlements, Lindiwe Sisulu, committed the government to the implementation of fifty catalytic megaprojects, each one of which would construct a minimum of 15,000 dwellings (Greve 2014).

Thus, unpacking some of the dynamics of this case may begin to bring to the surface some of the intersections and relations that occur in PPP-driven suburban development projects in the country.

Land Assembly and PPPs: Bringing the Concepts Together

There is significant variance in the definitions of what constitutes a PPP, but there is consensus that their roots are located in the zero-based budgeting and "privatization" drives of North America and Britain during the 1970s and have been closely allied with Thatcherite cutbacks in public expenditure and the "Reaganomics" of private-sector led development (Mitchell-Weaver and Manning 1991). The idea was to combine "the initiative of the private sector investor with the foresight and public mindedness of government" (ibid., 46). Brought to developing countries on the back of structural adjustment policies first in the agrarian sector and later for urban development, PPPs are seen largely as cooperative agreements between actors situated within both the state and the private sector, including civil society and "big capital." Mitchell-Weaver and Manning (1991, 48) argue that PPPs "are primarily a set of institutional relationships between the government and various actors in the private sector and civil society." Given the limited space, debates concerning the history (Payne 1999; Jones and Pisa 2000; Choe 2002) and contrasting definitions will not be rehearsed in detail here. Rather, we note the two main models on how PPPs can be implemented: one may be termed the corporatist/privatized model and the other may be seen as the collaborationist model. There are some commonalities: both see PPPs as interactions between at least two actors, one from the public and one from the private sector; both understand PPPs to entail enduring and long-standing relationships, where each party brings something of value and shares risk within the life-cycle of the project (Zou, Wang, and Fang 2008). However, there is significant disagreement in the detail of how each model operates.

The first, the corporatist model, sees PPPs as the method through which there is "extensive … privatizing facilities and services, or may be simply obtaining management or financing techniques from the private sector" (ibid., 123). In such a conceptualization the state takes a step back and becomes the purchaser or client and attempts to facilitate and ease the way of the private sector organization to which responsibility has been assigned (Li and Akintoye 2003). Critics state that such a model is not truly a PPP but rather some type of urban regime or third-party government (Peters 1987) that is mired in an unfettered, free market approach and where the state abdicates some of its responsibilities. Thus, the second model offers a different view of the elements that constitute a PPP, namely, "Goals and objectives can be set in common. Joint management and regulation can occur," where non-state parties such as private capital, civil society, and communities are brought into the decision-making process (Mitchell-Weaver and Manning 1991, 49). In the second model the state is an active regulator and participant throughout and does not abdicate responsibility. In addition, finance and financial risk is shared and is not just the disbursement of subsidies to private actors (Vickers and Yarrow 1988). In summary, the second model, and one that we use to analyse the following case, sees PPPs as having five key dimensions, underlain by a sense of shared and equal power between the participants. These dimensions include a strong and continuous presence of the state as regulator, engaged in the protection of all parties; joint goal-setting to satisfy the needs and requirements of all parties; joint financing and risk-taking; progressive governance in which all parties are able to "voice" their needs and contribute to the outcomes; and lastly, an enduring and long-term relationship between the various parties. Having identified the key elements or dimensions, we will engage with how the partnerships have been operationalized at each stage of the land assembly process.

We will also attempt to add a further layer of complexity to the analysis by considering the role of party politics and the use of informal channels of communication and access to those in power, which are often a feature of urban governance in cities of the global south (Chatterjee 2004; Bénit-Gbaffou 2008; Roy 2009).

This brings us to the next arena of conceptual terrain: that of urban land assembly. There are two ways of thinking about urban land assembly, one narrow and one broad. The narrow way sees it simply as the moment when land is acquired through private sector transactions, or even state intervention in the form of expropriation (Louw 2008), but it is the point at which there is a change of ownership. Generally, the transaction is accompanied by a transition from passive to active ownership, where the new owner is looking towards the development of the land in question (van der Krabben and Jacobs 2013). In contrast, the broad definition according to Golland (2003) sees land assembly as part of the development process as a whole. It could involve land acquisition from landowners; land preparation; planning of streets, open spaces, and main services; planning the built form; division of land into building plots; and delivery of the planned form.

For the purposes of this paper we view land assembly in the broader sense, including the transfer of ownership and transition from passive to active ownership, and also the various steps in development, readying land for new tenure formulations. Thus, in constructing a conceptual framework that is useful in analysing PPPs in the South Africa land development context, there are some adjustments that need to be made to the above-mentioned phases. Although land-use planning has recently changed with the promulgation of the Spatial and Land Use Management Act, we note several slightly different phases in the land-assembly process in Johannesburg, whereby agricultural land is converted to residential or mixed-use settlements known in South Africa as townships (it is important to note that all suburban settlements are known officially as townships in South Africa, not just low-income settlements as is commonly thought). Land development has to go through the following phases:

- The proposed township must undergo a set of environmental assessments.
- The existing stakeholders of the area must have a chance to take notice of and to provide inputs into the process when a landowner wishes to establish a township.
- The land must be provided with cadastral data in the form of a general layout plan, which designates stand sizes, numbers, land use, zonings, and tenure formulations.
- Once the layout plan is approved, the newly created properties must be registered in the deeds office.
- The land must receive services as required – both on the land itself and to be connected with the existing service system of the built environment, that is, roads, water, sewerage disposal, storm water outlets, electricity supply, and waste disposal; Installing these services is known as bulk infrastructure.
- The land will receive specific land-use rights both of which need to be incorporated into the existing town-planning scheme.
- Cash contributions will have to be made to the municipality regarding the extension and adjustments to the main service by the private developers; they are known as "bulk contributions."

For the most part, these phases are similar to the process of land assembly in other contexts. However, a fundamental difference, which brings back the notion of partnerships into the discussion, concerns the rights that illegal land occupants have in South Africa and the implications that they have for land assembly and PPPs. People living in these situations cannot be summarily evicted without certain conditions being met, including meaningful engagement and the provision of alternative accommodation if the eviction resulted in making households homeless. As such, land assembly in South Africa also has to include some formulation of the process of land sharing in which "an agreement between the illegal occupants of land and the landowner ... involves illegal occupants moving off high value land in return for being allowed to rent or buy part of the land below market value" (Ovens and Kitchin 2008, 15). Thus, illegal dwellers need to be included in the PPP as active members able to engage, as far as possible, in the dimensions discussed earlier.

The process is complicated. To disentangle the main elements we trace the development of Fleurhof by examining the nature of the partnerships at each phase of land assembly. Our purpose is to see how, or if at all, the dimensions of PPPs discussed are operationalized. We are attempting not to appraise or assess how well or badly the various partners have carried out these dimensions, but rather to understand how the various partners engaged and interacted at each phase.

The Context

The Fleurhof scheme is a mixed-income housing project that, when completed, will comprise over 10,000 units: one-third fully subsidized by the state, one-third partially subsidized by government, and one-third intended for private sector ownership (Table 11.1). The mix includes state-subsidized rental (or "social housing"), fully subsidized units, colloquially called RDPs or BNGs (fully subsidized units given with full ownership to qualifying beneficiaries, built under the Reconstruction and Development Program over the mid-1990s and under the 2004 National Housing Policy called Breaking New Ground), as well as homes that have been bought using state-guaranteed loans under the Financial Linked Income Subsidy Program (FLISP) housing and bonded units. It is one of the first fully mixed-income, mixed-use housing developments in the country. This development has what is, in the South African

Table 11.1. The composition of Fleurhof housing typologies

Type	Tenure	Typology	Number of units
RDP/BNG	Ownership	3–4-storey walk-ups 40 m²	2,904
Gap	Ownership	Mix of detached; semi-detached 116 m²–400 m²	2,124
FLISP	Ownership	Mix of detached; semi-detached 3–4-storey walk-ups	2,755
Social Housing	Subsidized rental	3–4-storey walk-ups	1,431Josh co 452 MHA 538
Private Rental	Market related rental	3–4-storey walk-ups	765I HS 162
Sectional Title	Ownership is not subsidized	3–4-storey walk-ups	210

Source: N. Erasmus (pers. comm. 2014)

context, a highly unusual combination of tenure forms and housing types as well as the range of amenities and facilities such as business centres, crèches, schools, and some retail areas that have been put into the newly developed suburb.

The site is located within the old mining belt on a main road that acts as a key east-west transport corridor cutting across the centre of Johannesburg and now traversed by the newly installed Rea Vaya Bus Rapid Transit System. It is 13 km from Johannesburg's original CBD, on a buffer strip between what was an independent "white" municipality of Roodepoort and Soweto, which is the largest and best known of Johannesburg's racially segregated townships. It is very close to a set of industries and some mine dumps. To the east and southeast of the project are older, pre-existing, low-middle-income Coloured communities (people of so-called mixed-race ancestry, who were defined as a separate "race group" under apartheid and now are a self-identifying social group, with specific history, dialect, and culture; see Adhikari 2005), who have been in the area for many decades (see Figure 11.1).

Fleurhof was developed between a large number of private sector and public actors, who entered the land-assembly process in different

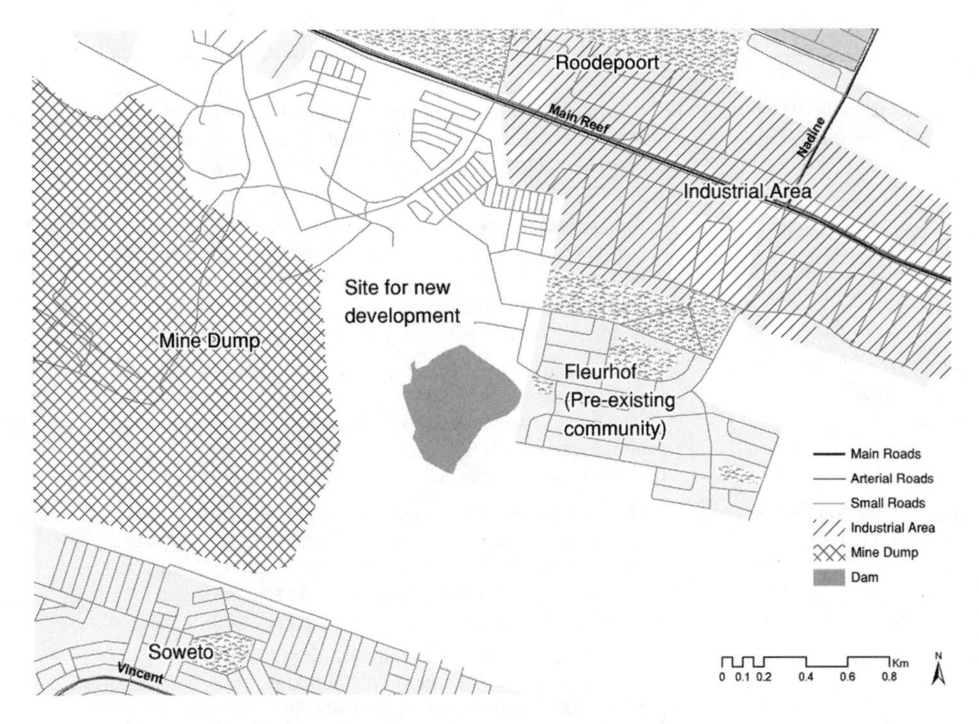

Figure 11.1 The situation of Fleurhof, South Africa. Sources: Calgro 2013; Selepe 2017. Copyright: Margot Rubin

configurations at different points. For convenience, and given the range and number of actors, Table 11.2 lists the dramatis personae.

The Dynamics of Partnerships in Land Assembly

In the previous section we described a number of stages to land assembly, but only the key elements, or rather an overview of the key moments of discussion and negotiation, are unpacked in the following section.

Starting Off: Land Acquisition

Land acquisition took both a conventional and an unconventional turn in the process. Formally, there was a transition from passive to active ownership; informally, there were a series of negotiations that allowed

Table 11.2. A list of key actors in the Fleurhof project

Key actors	
International Housing Solutions (IHS)	International Housing Solutions is a global private equity firm. It partners with financial institutions, real estate developers, and local government authorities to provide equity finance for housing projects (IHS website).
Calgro M3	A private sector development and construction company listed on the Johannesburg Stock Exchange, specializing in mixed-income residential projects. It has been in the market since 1995 and has considerable experience in all aspects of housing and housing programs (Calgro website).
First National Bank (FNB)	One of the "big four" commercial and retail banks in South Africa and one of the first to get involved in the "gap market" (those who cannot afford private sector provided property and do not qualify for state-subsidized housing.) (pers. comm., private sector banking rep., 2014).
Gauteng Partnership Fund (GPF)	GPF is a development finance institution that looks for ways of leveraging finance for affordable housing in Gauteng province. It was established in 2002 by the Gauteng Department of Housing and uses public and private sector funding to finance and gear affordable housing projects (GPF, website).
City of Johannesburg (CoJ)	The CoJ Metropolitan Municipality is the local authority that governs the municipality of Johannesburg. It comprises seven sub-regions and has an executive mayor and a system of ward representation consisting of 260 councillors (CoJ website).
Johannesburg Social Housing Company (JOSHCO)	JOSHCO is a municipal entity, wholly owned by the CoJ, to whom it reports. The entity is responsible for the provision and management of low-income rental housing for residents of the city and includes social housing, old hostels, as well as public-owned stock (JOSHCO website).

for land sharing and access to formal housing in the development by illegal dwellers who had been living on the site.

The conventional aspect of land acquisition began when the project was initiated by Calgro M3, a profit-driven private sector company that bought the land on which Fleurhof is now located. The mining company that had owned the parcel had done nothing with it and the site had stood mostly vacant and dormant for decades. Calgro M3 purchased the site in 2008, with the intention of going through the full township establishment procedure, rezoning the land from agricultural to residential and mixed use in order to generate a profit. Their 2008 annual report anticipated: "the estimated turnover from this project is R1.6 bn" (Calgro M3 2008). The total cost was projected to be R2.82 billion (about USD $200 million). Not having sufficient capital to buy the land, Calgro M3 entered into a partnership with First National Bank (FNB), one of the largest commercial banks in South Africa. First National Bank provided the mortgage finance, which allowed Calgro to buy the site and begin development.

The formal transfer of ownership was only one part of the land acquisition process; the second part was how the developers actually accessed land when people were already living there without formal tenure or official permission but nonetheless with some legal rights. A land-sharing negotiation took place between these parties, which allowed land transfer and development to proceed.

The site was undeveloped in 2009, except for three small communities living there: a group living in the old mine hostel, an informal settlement, and a community that had settled in an abandoned mine community hall and sports centre without permission of the owners. Through negotiation, with involvement of third parties, each group largely was able to get what it wanted. During these negotiations party politics was a key factor in determining how the various communities accessed the decision-making process and the final allocation process.

Of the three groups only the residents of the informal settlement successfully negotiated their rehousing with the developers. According to Ben Pierre Malherbe, the company CEO, although Calgro did have an eviction order, the informal dwellers were willing to be moved and to be rehoused in the new development, primarily because they had been on the housing waiting list for state subsidized housing since 1996. Furthermore, according to the city's housing official, it was also "because people have seen what these type of developments [PPPs] are. And people have started to embrace them." They were therefore quite happy to move. Allocation of units in this way is extremely unusual, and there are few precedents for such poorer communities to be made part of the negotiation and allocation processes in large PPPs. In the course of an interview David Dewes, the local Democratic Alliance (DA) councillor for the area (the DA is the major opposition party in South Africa), argued that it was because the informal dwellers were supporters of the African National Congress (ANC), the governing political party, that they were able to negotiate with the city and obtain RDP housing in Fleurhof.

Figure 11.2 The remainder of the original Fleurhof informal settlement, South Africa. Source: Margot Rubin, 2013

The other two communities were not part of the original negotiations, and Calgro tried to evict them. However, Dewes was able to represent their interests in negotiations with the developer. His actions were apparently because of the apathy of the City of Johannesburg, and the local ANC branch, which were well aware of the eviction but did not engage with the communities, presumably because they were too small to be of any use politically (around a few dozen households) (Dewes, pers. comm. 2014). Dewes was aware of the project from the beginning and had been engaging with the community since development began. Given his close relationship and consistent discussions with the residents, he knew what they wanted and was able to negotiate on their behalf. Dewes threatened the developers with bad press and told Calgro that, if they did evict, he would "call a press conference … and … get 30 journalists in here and … advertise that this [the eviction] wasn't done humanely" (ibid.). Calgro then became more open to discussion. Ultimately, the company agreed that the community would be transported to the informal settlement, have some form

Figure 11.3 Original mine hostel, Fleurhof, South Africa.
Source: Margot Rubin, 2013

of shelter rebuilt for them and eventually have access to RDP units as part of the project.

The other pre-existing community, the hostel dwellers, also were at first excluded from the relocation plan. Once again local politics was to blame, as the people in question were largely isiZulu, and supporters of the Inkhatha Freedom Party (IFP), which was in opposition. There are deep-seated tensions between IFP hostel dwellers and local ANC supporters, who often consider them to be usurpers, not legitimate residents, although many have lived in these areas for decades. In this case Dewes, who sincerely "wanted the best deal for everyone" (ibid.), ensured that the hostel dwellers were able to access housing in either RDP or state-subsidized rental units.

As a consequence of these negotiations and various political interventions, land-sharing agreements were put in place. They ensured that the poorer communities would move where and when requested and would not put any legal barriers in place or cause media embarrassment; in exchange, they were allocated state-subsidized units within the project or serviced sites in an associated project.

Land Preparation

Once land has been acquired it must be rezoned from agricultural to urban use. The partnerships and negotiations on this issue centred on a few key features: greasing the wheels of government; a negotiation on the layout of the settlement for the final plan; and discussion over the terms of bulk infrastructure provision.

The original layout plan of the settlement was designed by Calgro M3 but was revised after discussions with the City of Johannesburg and with community members living in the adjoining neighbourhoods. South African environmental and planning legislation requires engagement and consultation with affected communities. Such activities generally have little effect on the outcome of the project; the Fleurhof case was quite different. As mentioned, it was located next to an existing residential area. This suburb was composed mostly of a low-middle-income community that had lived there for decades. When first told about the new development, residents expressed significant concerns. They cited many of the usual middle-class anxieties around public housing developments, neatly summarized by Ros Morta, a twenty-year resident of the area:

> The general motion in Fleurhof is that we don't agree with where these houses are being built and we are worried about the possibility of a harsh decline in our property valuations.
>
> The neighbourhood has prided itself in being relatively crime-free. How can we manage that standard and our safety with these developments?"
>
> I cannot afford a possible security breach here. I have grandchildren and sons to worry about. (Quoted by Myburgh 2013)

Calgro and Dewes engaged with residents over the project, informing them of the plans and explaining the intention and outcomes. However, despite these reassurances, the project still faced considerable resistance. Nonetheless, Dewes (pers. comm. 2014) apparently told his constituents: "This [the Fleurhof development] is the reality. Government wants to densify ... You're not going to stop this. So, what we've got to do as a community is we've got to negotiate the best possible scenario." The residents then came to an agreement that they would ask the developer to construct some type of buffer between them and the RDP housing, in an attempt to protect their land values and their families from perceived threat (ibid.). Calgro M3 and the city acceded to this request; there is a small space and a set of bonded houses (i.e., houses for slightly higher-income groups who can access and afford private sector mortgages), which were built between the older, pre-existing settlement and the new development.

While the land was being purchased, Calgro was already in discussion with the city. The company needed it to come on board for several reasons. In order for the development to be profitable, they needed the city to pay for bulk infrastructure. According to Ben Pierre Malherbe, the Calgro CEO (pers. comm. 2014), they knew that "the city's involvement actually makes it possible to provide significant social housing because otherwise it is just too expensive, doesn't work. The fact that they have come in with municipal infrastructure grants (later called Urban Settlements Development Grant [USDG] funding) makes it doable from a developers' point of view."(USDG, which is administered through the National Treasury, seeks to support the development of sustainable human settlements and improved quality of life for households through accelerating the provision of serviced land with secure tenure for low-income households in the large urban areas by supplementing municipal resources). Malherbe noted that the company's opening gambit was to exchange their land and to build fully subsidized units, "in return for assistance with the bulk requirements [major infrastructure connecting the settlement to the network of services] for the town" (ibid.). However, the city had its own requirements and was prepared to consider the project and provide the necessary permission only if it contained mixed housing. A senior housing official said, "if you give me a proposal, that has, as an end project, [a] majority of units that will cater for people who earn nothing, people who earn R5 000 to R11 000 in terms of FLISP, and some social housing stock, we can talk" (pers. comm. 2014).

The city did not stop there and insisted that, in exchange for the infrastructure upgrades, support, and assistance, the developers should pay for additional amenities. These amenities included a new water reservoir, four business centres (mixed use), seven crèches, five religious sites, one community centre, three schools and twenty to thirty parks scattered across the settlement. To further their goal of "forced racial integration," the city also insisted, "we also want to do a double carriage that will link Dobsonville [a suburb in the historically Black township of Soweto] with Fleurhof, by default you link it with

Roodepoort [the traditionally White suburb to the north]" (ibid.). It was only if Calgro agreed to these conditions that the city stated it would apply for a grant for servicing the land. In this case they would apply for the USDG from National Treasury, without which the project would have been too expensive.

Calgro also needed the city to help navigate the labyrinthine corridors of power to gain planning permissions. Thus, Malherbe (pers. comm. 2014) stated, Calgro asked the city to apply for certain kinds of permissions on their behalf: "we [the developers] are hiding behind them [the CoJ] when it suits us, in our development agreements, so when we go to City Power, we need power, for 10 000 units, much easier getting it if the application's coming straight from the City of Joburg Department of Housing." Rob Wesselo, a managing partner with IHS, also recalled that the city agreed to assist in pushing through servicing agreements and land approvals and in doing so made the project quite "speedy" and reduced holding costs (pers. comm. 2014).

Thus, the layout plan that was originally designed was amended significantly by the inputs and requirements of both the surrounding community and the local government. These changes included the number and size of plots, whom they would be allocated to, and the form of tenure that would be in place. However, there was a point beyond which the developers said they would not have been willing to go; as a result, while the amendments possibly lowered some of their profits, the profit that was made was still well within the margins that they and their investors expected. The situation was not one of the developers holding all of the cards, and the CoJ negotiated quite hard to ensure that its requirements around settlement design met with policy requirements and political commitments. Once it had commitments from the developer, it was then willing and able to lubricate the wheels of bureaucracy and ensure that the relevant planning applications were expedited and bulk services were hooked up.

Planning the Built Form

The final form of Fleurhof, how it looked and whom it housed, was heavily influenced by the negotiations of the partners who entered into relationships with Calgro. The following section discusses the nature of these negotiations and how they shaped the architecture and built form of the settlement.

Once again, in exchange for assistance in gaining access to funding for bulk infrastructure and getting the requisite approvals, the city insisted that the design of the RDP units be changed, arguing, "the profit that you [the developer] are making when selling your FLISP [finance linked individual subsidy program] unit, and your bonded units, you need to take some of that money and cross-subsidize to provide a higher level of service on your RDPs" (housing official, pers. comm. 2014). In effect, the city demanded a higher-quality RDP unit within the settlement and thus an improved "look and feel" of the state-subsidized housing.

Furthermore, the two social housing institutions, Madulommoho Housing Association (MHA) and JOSHCO, which had been brought in owing to the city's social housing requirement, made inputs into the final housing design. JOSHCO is a municipally owned entity and is effectively the organ of the City of Johannesburg responsible for taking on and managing social housing, and MHA is a non-profit social housing institution; eventually, they compromised on a uniform design that satisfied both of their needs.

Shiraaz Lorgat, head of Development Strategy Consulting at the Gauteng Partnership Fund (GPF), which helped to finance the social housing portion of the project, remembered that MHA and JOSHCO worked together and agreed that they would supply "mirror developments" (pers. comm. 2014). All of their buildings would be of similar quality and amenities, for example, including heat pumps and access control. The idea was that it would be difficult to tell which units were JOSHCO's and which were MHA's. Neil Erasmus, chief operating officer of MHA, noted that although MHA could have charged higher rentals, they negotiated with JOSHCO and "agreed to keep it on par to keep the peace in the area and not start an uprising, because it is the same project. So definitely it was a calculated move to engage with them to keep it very similar" (pers. comm. 2014). This decision was largely due to previous experience of social housing around the country where tenants, on finding out that neighbours pay less or have better facilities, have been known to engage in rent boycotts or to leave, often crippling social housing institutions and endangering their sustainability. Lorgat also argues that it was about ensuring that the two institutions did not compete and could control their costs: "you don't want this competition between institutions so you don't want one guy jumping from here from there because it is, in terms of administration, it is [a] huge amount of work when somebody jumps from here to there so the institutions don't want that" (pers. comm. 2014).

MHA and Calgro were also able to discuss which buildings MHA would take over and what their needs were. Calgro was committed to providing social amenities but did not necessarily want to run all of them once the development phase of the project was completed, while MHA has a model, which includes taking substantial interest in residents' needs. Thus, MHA was able to get a building with a crèche on site, which suited both their needs and Calgro's. Furthermore, MHA was able to influence building specifications that are particular to social housing: "Sometimes you have to be more careful how you design for social housing than … for the market" (N. Erasmus, pers. comm. 2014). Because of the number of tenants that the units need to accommodate over time, more robust fittings are required. Thus, "once we [MHA] became acquainted with Calgro and we were able to sit around a table and really see eye to eye on how to meet each other's needs," they developed a "nice synergy" (ibid.).

Development Financing

A further element of land assembly that needs to be considered is the nature of development financing, which is usually crucial in every stage of the land-assembly process. As will be demonstrated, all parties, private and public, provided some form of funding for the project. It was also reported by a number of respondents that there was almost a chain reaction whereby, because government was on board, the private banks and financiers were happy, even eager, to get involved, as the state subsidies lessened the risk and helped to ensure profitability. Similarly, the state was willing to look for additional funding when it was apparent that there was private sector interest and investment. The phasing of financing occurred as follows. The land was initially bought in a partnership between Calgro M3 and First National Bank. Once the land was purchased, the state applied for Urban Settlement Development Grant funding to provide bulk services, after which equity was needed to begin construction. The global private equity firm IHS came forward and bought 30 per cent of the project for R50 million. They had undertaken extensive market research and found the gap market underserved and underserviced. They saw Fleurhof as a project that fit this niche. Bulk infrastructure – notably connections to the city's sewerage works, electricity, roads, and so on – and internal infrastructure costs, were financed by the state through USDG. Housing subsidies were negotiated by the City of Johannesburg and supplied by national and provincial governments (senior housing official, pers. comm. 2014). Calgro M3 brought together all of the partners and managed and oversaw the project (IHS website; GPF website; Malherbe, pers. comm. 2014). The GPF, the Social Housing Regulatory Authority (a state organization mandated with funding and supporting subsidized rental), and ABSA (Amalgamated Banks of South Africa), one of the "big four" banks in the country) provided funding for the social housing component of the project and some funding for some of the sectional title/private units that were developed. GPF's total exposure was just under R55 million.

Urban Management: Continued Engagement

Although not strictly a part of land assembly, urban management may be seen as a form of continued engagement, which is an element of a PPP. Since the initial funding phase IHS has chosen to remain involved and now owns 162 sectional title units, which it rents in the commercial market. Of the 990 social housing units that were built, JOSHCO, as the city's municipal entity in charge of social housing has taken over management of 400 units. MHA has bought the remainder of the social housing units and is managing them. All of the banks, rental, and social-housing institutions have reported enormous demand for all of the housing units and there are also people on site who act as real estate agents (for Calgro) (JOSHCO website; T. Erasmus, pers. comm. 2014). The banks also insisted that Calgro continue to be involved and has agreed to sit on the body corporates of the sectional title buildings in order to ensure good governance of their investments. The CoJ has also taken on some of the responsibility for urban management and maintaining the area.

Conclusions

The case study of Fleurhof brings to the surface some insights into Public Private Partnerships in South Africa, and indeed beyond. The first is that power is not necessarily asymmetrical, as one might initially assume. In this case none of the partners was able to absolutely dictate the outcome of the partnership. The city used its ability to request funding for infrastructure development and navigate the land preparation application system as a form of leverage to meet its political and policy requirements. Calgro M3 and its funders were able to ensure their profitability by guaranteeing delivery of houses for the very poor. Through the intervention of local political leaders, the local communities were also able to have input into the process and the final layout and allocation of the development, although the situation does raise the question of whether a similar outcome would have occurred had there been a less effective or less interested political agent.

It would also appear that current planning and tenure regulations have had a direct effect on the ability of communities and poorer people to negotiate, ensuring that, rather than straight eviction, forms of land sharing have had to be considered and included. The land-assembly process has also meant that there were a number of points at which partnerships could occur and in a sense encourage joint ventures between a number of public and private sector players.

Also important were long-standing relationships between the various partners whereby the history of interaction between MHA and JOSHCO meant that both parties were open to negotiate. The previous experiences of the City of Johannesburg with Calgro on other projects meant that there was also a relationship of trust on which to build, which made partnerships easier. However, the promise of future relationships also seems to have played a role: the city and Calgro hope to do more projects together in the future; as do IHS and Calgro, which have very similar plans for developing more housing in the lower-income market in the future; as do the commercial banks, which are keen to maintain a presence in this market. Mutual self-interest also seems to have lubricated negotiations and ensured that each party was reasonable in its demands.

It would appear that at every point in the land assembly process there were significant negotiations and joint decision-making, which had tangible outcomes for the final spatial layout and built form. The PPP also usefully ensured that poorer people gained access to good-quality housing and added significant amounts of stock to the lower end of the housing market. Thus, there is much to suggest that in the case of Fleurhof the PPP has had material benefits for all concerned and seems promising as a form of land assembly and urban development in the future.

APPENDIX

List of Interviewees

Name	Organization	Date of interview
Lael Bethlehem	Former CoJ official and former head of Affordable Housing	12 June 2013
Anonymous 1	Senior asset manager, IHS	11 July 2013
Gina Zanti	Director, Development Planning, City of Johannesburg	01 August 2013
Shiraaz Lorgat	Head, Development Strategy Consulting; Gauteng Partnership Fund	28 January 2014
Neil Erasmus	Chief operating officer MHA	06 February 2014
Ben Pierre Malherbe	Chief executive officer, Calgro M3	25 March 2014
Rob Wesselo	Managing partner, IHS	26 March 2014
Charles Davis	Mixed-income housing Lufhereng, City of Johannesburg	26 March 2014
Housing Official	Senior housing official, City of Johannesburg	15 April 2014
David Dewes	Councillor for Fleurhof, Democratic Alliance	22 April 2014
Anonymous 2	Senior representative of commercial bank	23 April 2014
Tinus Erasmus	Project manager, Calgro M3	Telephone conversations and email

REFERENCES

Adhikari, M. 2005. *Not White Enough, Not Black Enough: Racial Identity in The South African Coloured Community*. Athens: Ohio University Press.

Bénit-Gbaffou, C. 2008. "Are Practices of Local Participation Sidelining the Institutional Participatory Channels? Reflections from Johannesburg." Transformation, Critical Perspectives on Southern Africa 66 (1): 1–33. https://doi.org/10.1353/trn.0.0003.

Calgro M3. 2008. *Calgro M3 Holdings Limited – Audited Annual Results for the Year Ended 29 February 2008*. Release date: 26 May.

Calgro M3. 2013. "PPPs … Competitive Advantage or Curse? A Practical Guide to Effective Relationship Building." Presentation at the South African Housing Foundation Annual Conference, Cape Town. http://www.sahf.org.za/Images/2013%20Proceedings/Power%20Points/MALHERBE,%20BEN%20PIERRE.pdf.

Chatterjee, P. 2004. *The Politics of the Governed – Reflections on Popular Politics in Most of the World*. New York: Columbia University Press.

Choe, S.C. 2002. "The Promise and Pitfalls of Public-Private Partnerships in Korea." International Social Science Journal 54 (172): 253–9. https://doi.org/10.1111/1468-2451.00377.

Golland, A. 2003. *Models for Land Assembly in the UK: A Comparative Analysis of Other European Approaches*. London: RICS Foundation.

Greve, N. 2014. "Govt Megaprojects to See Construction of 1.5m Houses in 5 Years." *Sisulu, Engineering News*. 16 October. Accessed 19 February 2015. http://www.engineeringnews.co.za/article.php?a_id=349049&rep_id=4136.

Jones, G.A., and R.A. Pisa. 2000. "Public-Private Partnerships for Urban Land Development in Mexico: A Victory for Hope Versus Expectation?" Habitat International 24 (1): 1–18. https://doi.org/10.1016/S0197-3975(99)00024-7.

Li, B., and A. Akintoye. 2003. "An Overview of Public-Private Partnership." In *Public-Private Partnerships: Managing Risks and Opportunities*, ed. A. Akintoye, 3–30. Oxford: Blackwell. https://doi.org/10.1002/9780470690703.ch1.

Louw, E. 2008. "Land Assembly for Urban Transformation: The Case of s-Hertogenbosch in The Netherlands." Land Use Policy 25 (1): 69–80. https://doi.org/10.1016/j.landusepol.2006.09.002.

Mitchell-Weaver, C., and B. Manning. 1991. "Public-Private Partnerships in Third World Development: A Conceptual Overview." Studies in Comparative International Development 26 (4): 45–67. https://doi.org/10.1007/BF02743762.

Myburgh, M. 2013. "Fleurhof Concerned by Latest Developments." 11 July. Accessed 13 May 13, 2014. http://biz.hozi.co.za/.

Ovens, W., and F. Kitchin. 2008. *The Recognition and Enhancement of Socially Dominated Urban Land Markets. Urban Landmark Report on Municipal Workshops and Proposal Selection, Urban LandMark*. Accessed 20 February 2015. http://www.urbanlandmark.org.za/downloads/Recognition_Municipal_Workshops.pdf

Payne, G. 1999. "Public/Private Partnerships in Land for Housing." In *Making Common Ground: Public/Private Partnerships in the Provision of Land for Housing*, ed. G. Payne, 1–16. London: Intermediate Technology Publications.

Peters, G. 1987. *Third-party Governments and Public-Private Partnerships.* Department of Political Science, University of Pittsburgh, M5.

Roy, A. 2009. "Why India Cannot Plan Its Cities: Informality, Insurgence and the Idiom of Urbanization." Planning Theory 8 (1): 76–87. https://doi.org/10.1177/1473095208099299.

United Nations. 1993. *Public/Private Partnerships in Enabling Shelter Strategies.* Nairobi: United Nations Centre for Human Settlements.

van der Krabben, E., and H.M. Jacobs. 2013. "Public Land Development as a Strategic Tool for Redevelopment: Reflections on the Dutch Experience." Land Use Policy 30 (1): 774–83. https://doi.org/10.1016/j.landusepol.2012.06.002.

Vickers, John, and George Yarrow. 1988. *Privatization: An Economic Analysis.* Cambridge, MA: MIT Press.

Zou, P.X.W., S. Wang, and D.P. Fang. 2008. "A Life-Cycle Risk Management Framework for PPP Infrastructure Projects." Journal of Financial Management of Property and Construction 13 (2): 123–42. https://doi.org/10.1108/13664380810898131.

chapter twelve

Production of Land for Real Estate Markets in the Suburban Area of Chennai Metropolis: The Case of Sriperumbudur-Oragadam Region

BHUVANESWARI RAMAN

The Sriperumbudur-Oragadam suburban region on the outskirts of Chennai metropolis in the South Indian State of Tamil Nadu is projected as a new growth centre and as a favoured destination for real estate investment (Ernst and Young 2014; JLL 2014). The government of Tamil Nadu intends to develop the region as a global hub of manufacturing. It implemented several projects for building industrial parks (IPs), Special Economic Zones (SEZs), and Industrial Corridors (ICs). These interventions accelerated the real estate development in the region and put upward pressure on land prices. Indian and international real estate development firms promoted integrated townships and gated housing complexes, anticipating the demand for housing from the employees of multinational companies and non-resident Indians. This chapter explores the micro-practices of land transformation in the region.

Private developers are reconfiguring the suburban landscape of Indian metropolitan regions and large cities (Shatkin 2011; Goldman 2011b). They are partnering with government agencies to build large infrastructure and housing complexes. India's government promotes Public Private Partnerships (PPPs) and the integrated township as a blueprint for urban development (Shatkin 2011; Chatterjee 2013; Sharma, 2013). Further, both the national and the regional governments in India introduced new laws, planning instruments, and institutions to create a favourable environment for private developers (Benjamin and Raman 2011; Datta 2015; Dutta 2012; Shatkin 2011; Goldman 2011a, 2011b).

Developers with transnational networks increasingly influence urban development agendas (Searle 2010, 2014; Rouanet and Halbert 2016). They anchor transnational capital (Searle 2010; Halbert and Rouanet 2014) and leverage their financial power to reform laws and institutions (Searle 2010, 2014; Goldman 2011a; Rouanet and Halbert 2016). They orchestrated reforms on several fronts: repealing the Urban Land Ceiling and Regulation Act (ULCRA), digitizing land titles, and formulating a real estate investment bill (Searle 2010). Existing studies expose the practices of transnationally networked developers and their relationships with foreign investors. However, research on micro-practices of real estate development is scarce.

Real estate developers assemble resources such as land, finance, labour, and building materials, and negotiate state laws for implementing their projects (Searle 2010). In post-Independence India, private developers have coexisted with cooperatives and self-built and government-produced systems of building development (Baken 2003). Their scale of land development was relatively small compared with those developers promoting integrated townships and gated housing. Several questions arise, given the changing landscape of real estate development. What is the role of government agencies? Have the practices of land developers changed in the suburban and rural areas and, if so, how? Who are the developers? What types of project do they promote? How do they mobilize land and finance? Some of these questions are addressed here, drawing on the findings of a qualitative research conducted in villages in the Sriperumbudur-Oragadam region.

Different actors and government agencies are involved the process of converting agricultural land for urban use in various ways. First, regional governments, particularly industrial parastatals, are dominant actors driving the process. Second, private developers are not a unified category. Their activities differ in terms of the types of project, the scale of land development, and the practices of mobilizing resources, especially land and finance. Developers promoting large-scale residential projects, henceforth referred to as large developers, are one category of actor in the region. Several other actors, including landowners, are implementing projects of varying scale ranging from small to medium, hereafter referred to as small and medium developers. Since land transactions are embedded in sociopolitical relations, the process of land transformation consists of two activities, namely, land consolidation and land development. Land consolidation is a complex process, as landowners in the research context predominantly have small holdings. Large developers from India and abroad with weak ties to local landowners rely on locally embedded actors, including small developers or government agencies, to consolidate large land parcels for their projects. Land assemblers, many of whom have a small to medium scale of operation, rely on their social and spatial networks to negotiate with landowners. They sell either assembled large land parcels to large developers or small plots to retail investors. Third, landowners respond in complex ways to the state's land acquisition drive. Finally, different factors influence their decision to part with land, one of which is land acquisition.

Similar to peri-urban development in Bangalore, which led Goldman (2011a) to develop his theory of speculative urbanism, the speculative logic of a real estate economy is fuelling land transformation in the Sriperumbudur-Orgadam region. This case study adds to speculative urbanism theory by providing a nuanced account of the politics of transformation and the practices of different categories of government agency, private developer, and farmer. The entry of large investors did not dis-embed small developers from the region, but rather re-embedded them in new realities. Paradoxically, the very speculative logic that fuelled large developers' projects destabilized their expansion in the region. Both large and small developers readjusted their products to align with the changing political-economic milieu.

Practices of Worlding Cities

Goldman's (2011a) Speculative Urbanism theory is used mainly to explain the forces driving the process of land transformation in suburban localities in India. Led by an alliance of government agencies in charge of urban development; large developers; and a transnational network of financiers, international development agencies, and consultants, the process is built on the assumption of increased demand for land transformation due to the demand generated by the putative expansion of corporate firms catering to global markets. Their strategies for land and infrastructure development hinges on debt-financing the construction and debt-servicing through tapping real estate gains, which in turn propels government agencies to continuously acquire agricultural land from small and medium farmers. Goldman (2011a) cautions that land dispossession has become a routine activity of government agencies; this has pushed inhabitants into a state of anxiety about losing their land to future projects. While speculative urbanism theory is useful to explain the logic of land transformation in suburban areas and its consequences for the city, it is difficult to apply to the micro-process of land transformation.

Other studies show how the political dynamics of land transformation are not straightforward. In some instances landowners have destabilized the government agencies' efforts to acquire their land and have even stalled large urban development projects (Datta 2015; Benjamin and Raman 2011). In some cities landowners have voluntarily sold their land to developers or formed partnerships with experts to build private townships in order to benefit from the increase in land prices (Vijayabaskar 2010, 2013; Balakrishnan 2013; Sami 2013; Kennedy et al. 2014). Moreover, private developers' projects for building integrated townships often remain on paper (Shatkin 2011). Similarly, Searle (2010) illustrated the manner in which globalization of real estate influenced the practices of India's large real estate development firms and their fragile relationship with foreign investors. Shatkin (2011) suggests further research on how the strategies of and relationships between the state, developers, and landowners influence the project outcome.

This contribution adds to Speculative Urbanism theory by providing a nuanced analysis of the micropolitics of transformation. It maps the practices of different actors involved in the production of land for suburban real estate markets.

The Case Study

This case study draws on research findings on the dynamics of land transformation in thirteen villages situated along two highways in the Sriperumbudur-Oragadam region. The Government of Tamil Nadu is developing highways as industrial corridors. The highways connect Chennai metropolis with Bangalore city in the neighbouring State of Karnataka and Mumbai metropolis in the Western India state of Maharastra. Bangalore is known for its global information technology industries, and Mumbai is the financial capital of India.

Several actors, including landowners, real estate developers, and government agencies, assemble and/or develop land in the thirteen villages. Private actors may be differentiated into three broad categories based on the scale of their projects. Large developers predominantly promote projects in an area of 50 acres or more; mid-level developers normally work within an area ranging between 10 to 30 acres; and small developers assemble land in an area of less than 5 acres.

Figure 12.1 New construction in Oragadam village, India.
Source: Image capture by Bhuvaneswari Raman, 8 July 2017

Different government agencies influence the process of land assembly and land development in the study area, including the parastatal agencies responsible for industrial development, the Department of Town and Country Planning (DTCP) and rural local governments. In addition, the Japan International Agency (JICA) and the Tamil Nadu Road Infrastructure Development Corporation (TNRIDC), a parastatal agency, are involved in planning, financing, and building road networks.

From 2005 onward the international media spoke of the ever-expanding real estate market and even called it India's Shenzhen (Ernst and Young 2014). However, almost ten years later, the scenario had changed, with the closure of manufacturing units like Nokia and Sybil, as well as a slump in the real estate market.

The field research for this case study is based on ethnographic work in thirteen villages in the Sriperumbudur-Oragadam region (eight villages with state-acquired land and five villages where private developers are leading the transformation). Designed in two phases, the project started with a rapid survey of the villages and interviews with field officials of the revenue administration, local political leaders, local government representatives, and large farmers in each village. The rapid survey and interviews were critical for insight into the broad patterns of land transformation and to collect basic information on the number of households, caste composition, land area, categories of land, area of land acquired or notified by the industrial parastatals, details of real estate agencies and their projects, history of land sales, and anecdotes of land conflicts. In the second phase we interviewed sixty farmers from different castes, landless households and tenant farmers, real estate developers (five with a small to medium scale of development and two large developers), bureaucrats, local and regional political leaders, Non-Government Organizations (NGOs), and political party workers. We analysed newspaper articles and internet sources in English to infer the trends in real estate markets.

The Role of Government Agencies

Different government agencies at the regional and the local levels influenced the process of producing land for real estate markets in the Sriperumbudur-Oragadam region. However, the regional government agencies have a greater influence over the process. The two industrial parastatal agencies of the regional government, namely, the State Industries Promotion Corporation of Tamil Nadu (SIPCOT) and the Tamil Nadu Industrial Development Corporation (TIDCO), are active in land assembly and development in the study area. SIPCOT and TIDCO are parastatal agencies headed by a bureaucrat and were originally established to promote small and medium industries in the city. SIPCOT notified for land acquisition via GO no. 258 on 27 November1995, no. 61 for Irungattukottai SEZ and for Sriperumbudur SEZ in 1997, and GO no. 125 for the Oragadam Special Economic SEZ on 9 May 1997. Until the mid-1990s SIPCOT acquired land from farmers and allotted developed land to small and medium enterprises at a subsidized rate. TIDCO was created in the early 2000s to develop land in partnership with private developers. The two agencies formulated several projects in the Sriperumbudur-Oragadam region, setting up SEZs and IPs in order to attract globally connected corporate enterprises. SIPCOT acquired land from eight villages in the study area to develop IPs and an SEZ, and TIDCO partnered with a private developer to build a township at Sriperumbudur.

The local governments in the study area have very little influence over the land development process. The production of land for real estate markets in the suburban areas is regulated by two regional government departments, namely, the Revenue Department and the Department of Town and Country Planning (DTCP). The former agency regulates the conversion of agricultural land to non-agricultural use, and the DTCP issues planning permits for land development. The local governments' role is limited to issuing no-objection certificates for land use conversion or land development. However, the elected representatives of rural local governments aligned with landowners in resisting SIPCOT's plan for acquiring land in the study area. Further, local-level political leaders are sought by both landowners and buyers to mediate land transactions.

Land and Housing Development: Actors and Practices

This section describes the practices of different actors involved in the production of land for the suburban real estate market. It shows that, in addition to large developers, small and medium developers shape the transformation of agricultural land in the Sriperumbudur-Oragadam region. Further, this section illustrates landowners' responses to the state's land-acquisition process, including a repertoire of actions such as protest, legal contestation, sale, and negotiation. Findings discussed in this section help to fill the gap in the Speculative Urbanism theory regarding the role of small and medium developers and landowners' complex responses to land acquisition.

Characteristics of Developers and Projects

Large developers' firms from India and abroad (United States, Europe, and the Middle East) have a presence in the Sriperumbudur-Oragadam region. These firms include Inno-Globe (the Indian subsidiary of ETA Holdings headquartered in the Emirates), Lancor (a private equity company founded by expatriate Indians of Dutch origin), Hiranandani builders from Mumbai, and Arun Excello and Marg, both of which are based in Chennai metropolis). Their projects are located on land parcels along the main highways and in the SEZs in the Sriperumbudur-Oragadam region. They predominantly build integrated townships, luxury apartments, and gated housing complexes, as can be seen in Table 12.1.

ETA Star India and Inno Group Holdings floated projects in the Sriperumbudur-Oragadam region after 2005. ETA Star-India is part of the multibillion dollar business conglomerate ETA Ascon Group. The group was established in 1973 as a joint venture between Tamil Muslim entrepreneurs and the UAE's Al Ghurair Group (Reuters 2007). Inno Group Holdings started as a private equity investment firm and subsequently entered property development. It is a partnership firm of

Table 12.1. Residential projects promoted by large development firms, India

Company	Type of development	Land area	Land assembly and type of development	Financial source
Inno Group Holdings	Inno Global Geo-city Township (Integrated township)	131 acres	Purchased land previously occupied by an institution and agricultural land.	Capital markets listed in NIFTY.
ETA STAR	Globeville Township	350 acres (of which 82 acres are allotted for housing)	Public-private partnership with TIDCO	Equity from Middle East investors, national banks, and retail investors
Arun Excello	Townships in Sriperumbudur and Oragadam	200 acres (for three townships approved by Department of Town and Country Planning)	Incremental investment over a period of more than ten years + purchase of assembled land	Capital market and regional banks
Lancor	Gated residential complex, Sriperumbudur. Apartments (Town and Country + Townsville), Sriperumbudur and Oragadam.	147 acres + 2 Acres	Land purchase directly from land holder (2 acres) + state allocation and purchase of assembled land from private investor in outright and joint venture basis (197 acres)	Capital market and regional banks
Hiranandani palace gardens	Oragadam	200 acres	Land purchased through market and Chennai Metropolitan Development Authority approved layout. Company has a Land bank.	Local banks (UCO bank) and HIRCO plc floated on London Stock Exchange
Marg	Gated residential complex (1800 apartments)	16.5 acres	Land purchased in market and DTCP approved layout	Local banks. listed on Bombay Stock Exchange

Source: Compiled from interviews with developers and internet research

Belgians of Indian origin and its projects are financed by non-resident Indians. Until 2012 Inno Group Holdings raised nearly $65 million for investment in Indian real estate (Lancor has been awarded a six-star rating by the CRISIL Real Estate Rating). Besides real estate, both ETA Star and Inno Group Holdings have insurance and infrastructure investment businesses. The Hiranandani builders and Lancor are Indian real estate development firms promoting projects in different cities, while Arun Excello and Marg properties are headquartered in Chennai. The latters' projects are limited to Chennai and other large cities in Tamil Nadu. The Indian companies are listed on the Bombay stock exchange and the National Stock Exchange. The Hiranandani builders also once controlled an offshore company that mobilized transnational capital.

Real estate development firms from Chennai metropolis and the neighbouring towns of Chingelpet have developed plots for residential use. Their projects cover a land area of between 1 and 30 acres, and are located between 5 to 20 km away from the main highway. They market their projects through advertisements on television, through digital media, and by word of mouth. Few landowners in partnership with real estate development firms have subdivided their farmland to form residential layouts. While the projects of large and mid-level developers' projects have DTCP approval, those of small developers may or may not have this approval.

Practices of Large Developers

Assembling land for township projects involves influencing farmers to sell their land and negotiating the prices, which cost both time and money. As can be inferred from Table 12.1, the township projects require assembling a minimum land area of 100 acres, which if purchased directly from landowners involves negotiations with many owners. A majority of landowners owned agricultural land between 1 and 5 acres. Very few households had land holdings of 10 acres or more. One or two agricultural land owners or real estate firms in four of the twelve villages researched held a maximum of 30 acres.

Large developers access land for their township projects through the state and the market process. In the former case developers enter into partnership with the industrial parastatal agency. For example, ETA Star entered into partnership with TIDCO to build an integrated township at Sriperumbudur. Interviews with a project manager and site engineers from ETA Star Holdings suggest that the company's managing director's close ties with the then chief minister were a factor in securing the PPP agreement. TIDCO committed 350 acres of land on a ninety-year lease, of which it transferred 82 acres. The parastatal facilitated speedy clearance of planning permissions. ETA Star prepared the township's master plan, which was approved by the Department of Town and Country Planning. The construction of residential apartments started in 2006, but was still incomplete during our field research in 2013. According to the company's executives, it had sold nearly 40 per cent of the flats as of June 2013. The company secured finance for construction from retail investors in the Middle East and Indian banks. In forging a partnership with the regional government, developers like ETA Holdings rely on networking with the regional political parties.

Large developers prefer large land parcels. Interviews with landowners revealed that they target the land of closed educational institutions and declining small industries or purchase land from small real estate development firms. An example is the Inno Global township project planned on erstwhile college-held land. The concerned college entrepreneur, a native of Sriperumbudur, invested in the region in the early 1980s, nearly 150 acres of agricultural land spread across eight villages, drawing on the support of the Naidu caste networks in local politics. A few large firms, such as Lancor Holdings, floated joint ventures with landowners.

Unlike the large real estate development firms from outside Tamil Nadu state, discussed earlier, firms from Tamil Nadu with extensive local networks engage in land assembly. For example, interviews revealed that a firm named Arun Excello, started by an entrepreneur from Chennai metropolis, undertakes land assembly. This firm promoted the first integrated township sprawling over 100 acres near Oragadam SEZ, for which it incrementally assembled land between 1980 and 1988. The firm purchased a cent of land for a price of Rs. 80–400, or a ft^2 of land for a price of Rs. 0.18–0.91, and sold built property at around Rs. 3,200 per ft^2 in 2012 (1 cent = 435.6 ft^2 or 40.47 m^2). Interviews with farmers at Oragadam suggested that dry land usually sold at Rs. 8,000 per acre and wetland at Rs. 20,000 per acre in 1980. Similarly, Marg Swarnabhoomi, another firm from Chennai, purchased agricultural land by mobilizing their Naikkar caste networks. In short, real estate development firms from outside Chennai or international firms rarely engage in land assembly. They draw on the support of their networks in the state agencies or political parties to access land via the SEZ projects or the market. Thus, it is not surprising that as early as 2004 the Confederation of Indian Industries (CII) at their annual conference demanded that the state amend its land acquisition act to quell opposition (Benjamin and Raman 2011; Chatterjee 2013). The owners of both firms – Arun Excello and Marg Swarnabhomi – drew on their caste networks in the Sriperumbudur-Oragadam region to secure information on land and to negotiate with the landowners.

Large developers use foreign institutional investment as a way to mobilize credit finance from the national capital markets and banks. ETA Star mobilizes finance largely from the Middle East, while Inno Global works with retail investors in the Netherlands and Belgium (Nandy 2012). Such capital may be used to purchase land for their land banks. According to a financial expert – an ex-employee of Price Waterhouse Coopers, now with the Indian Financial Service – developers largely use foreign institutional investments to raise investment capital from the Indian banks or the national capital market. It is often difficult to isolate the flow of money from foreign institutional investment and national sources into a project on the company's balance sheet, which is available in the public domain. However, a report produced by India's central bank shows that the real estate sector is a dominant lending portfolio for several local banks, making up around 25 per cent of the total loan portfolio in 2014 (Roy 2014). The CRISIL report on the real estate market in 2015 cautioned about the debt burden of the country's top twenty-five real estate development firms.

Several projects listed in Table 12.1 are far from completion. When the industrial parastatals announced their plan for setting up SEZs and IPs in the Sriperumbudur-Oragadam region, large real estate development firms from within India and abroad floated several residential township projects and gated housing complexes based on an assumed residential demand from the companies' senior executives and the Indian diaspora. Contrary to their expectation, completed township projects and gated housing complexes have very low occupancy rates, often only 30–40 per cent. The executives and white-collar professionals working at the SEZs predominantly live in Chennai and commute every day to Sriperumbudur. Moreover, non-resident Indians who invested in the much-hyped integrated township project – the Hiranandani Palace Gardens at Oragadam – face a risk of losing their investment. The project was stalled following a conflict between international investors and the Hiranandani builders and the closure of their financing arm, HIRCO plc. As the builders have defaulted on their payments, the banker has attached the Hiranandani Palace property in 2014. The Hiranandani case epitomizes the risks and uncertainty generated by the practices of large developers (Kamath 2014).

Large developers' projects are built on the assumption of an uninterrupted growth of global manufacturing companies, which in turn would stimulate the demand for real estate development. Their projects are financed through debts from both Indian banks and international capital markets, as can be inferred from Table 12.1. Their debt-servicing strategies rely on continuous growth. By contrast, some of the manufacturing companies have closed their operations and moved out of the region. In addition, the real estate markets have declined. The real estate shares of all major corporate developers have dropped steadily since 2011. Few large developers are retrofitting the apartments for sale to middle-income households. Despite these trends, industrial parastatals, especially SIPCOT, have announced further land acquisition to build more SEZs and townships. They have notified both agricultural land and small- to medium-scale residential layouts, which have been approved by the Department of Town and Country Planning (see Thangamani 2014).

The very speculative logic driving large development projects has disrupted their expansion. Thus far, large developers have not managed to eliminate small and medium developers. That said, the jury is still out. Recent laws introduced by the Government of India to regulate real estate and acquire land for development favour the interests of large developers. Large developers, along with international consultancy firms, lobbied to include a clause for issuing licences and restrict the role of small developers under the real estate regulation bill recently implemented by the Government of India. Further, the bill paves the way for easy liquidity of investment by foreign investors and was introduced to facilitate acquisition of land for large development projects.

Practices of Small and Medium Developers

Small and medium developers incrementally assemble land either on their own or in partnership with a farmer. Information gleaned from the rapid surveys and detailed interviews with agricultural landowners shows that farmers have gradually sold part or all of their land holdings since the late 1970s. Land sales increased in the 1980s and surged after 1998. Information on land prices collated from interviews with elected representatives of rural local governments and real estate investment

Figure 12.2 Transformation within the village in front of Hiranandani, India. Source: Image capture by Bhuvaneswari Raman, 9 July 2017

firms indicates a wide variation across the thirteen villages. Prices for a cent of land ranged from a low of Rs. 30 to a maximum of Rs. 80–100 in 1983; prices escalated to Rs. 1,000–4,000 between 1990 and 1998, and increased further to Rs. 6,000–10,000 in 2008. Land sales declined after 2011, but the price remained high in the range of Rs. 11,000–13,000 in 2013. This result was corroborated by the information provided by farmers on sale prices.

The developers pay landowners in instalments and rely on their investors' not defaulting on the payment. For example, a developer from Chennai incrementally assembled an area of 100 acres for the construction of a gated housing complex at Oragadam. The developer had invested in the villages around Oragadam since 1983. Another example is the practice of a mid-sized real estate development firm, the Awami Agency (AA). The agency, headquartered in Chennai, has been investing in Sriperumbudur-Oragadam area since 1973. The company promotes residential layouts – predominantly open plots and with a minimal construction component. The plots and bungalows are for a middle-class clientele from Chennai aspiring to own an independent house. AA is one of several such agencies found in the thirteen villages. The mid-size development firm located in Chennai or Chingelpet runs a regular transport service to bring potential buyers to their project sites in the Sriperumbudur-Oragadam region. Interviews with developers and village administrative officers suggest that disputes between land developers and farmers over land price increased after 1998. Some farmers refused to transfer the title deed and demanded a higher price after the conclusion of sale agreement.

The directors of real estate development companies often have previous experience in developing land and have worked or served as partners before starting their own company. Irrespective of the scale of operation, developers from Chennai and Chingelpet have been investing in the region since the early 1980s. Developers with a varying scale of operation realized their profits from the land price escalation post-2000.

Some small and mid-sized real estate development firms moved to land assembly after 2000 because of the possibility of securing high profits without investing in land development. According to developers, they face difficulties in getting panchayat permission. The case of a developer from the neighbouring small town of Villupuram is an example. The developer's firm is one the three firms operating in the villages close to Sriperumbudur industrial park. Until 2005 the firm developed residential layouts, which were approved by the Department of Town and Country Planning. The firm's owner purchased land from farmers and negotiated the transactions through his caste networks with Naikkar landowners and place networks with Naidu farmers. Large developers, without such ties, have found it difficult to negotiate with farmers at the local level. Another example is a developer from Chennai who incrementally assembled 70 acres between 1999 and 2003. According to a land developer, holdings change hands three to four times before being consolidated into a large parcel of land. The large land parcels are in great demand among real estate developers promoting luxury housing projects.

Two trends are found in the researched villages post-2014. On the one hand, there are several vacant units in the gated housing complexes and integrated townships, and construction of some township projects has stopped midway. On the other hand, the residential layouts formed by small to mid-sized development firms continue to grow despite SIPCOT's announcement about acquiring more land. The entry of corporate developers in the region has altered the practices of small and mid-sized real estate development firms, but the trajectory of change is fluid.

Practices of Agricultural Land Owners

Landowners responded in different ways to the parastatal agency's efforts to acquire land and to selling their land to private developers. Conflict over land acquisition is one dimension of their relationship with the state. Landowners across different villages formed alliances with retired bureaucrats and elected representatives of the village and local governments to contest land acquisition.

According to interviews with farmers and panchayat officials, SIPCOT has notified farmers about land acquisition on three different occasions since 1996. According to farmers whose land was acquired for the first industrial park, rumours about SIPCOT's plan circulated in the villages as early as 1995. SIPCOT demanded the farmers to surrender their land voluntarily. Farmers in one village in the first round gave up their land without much protest. However, farmers in several other villages did not respond to SIPCOT's demand. Senior leaders of the ruling party at the time, the Dravida Munnetra Kazhagam (DMK), intervened to increase the compensation amount. According to interviews with gram village panchayat (rural local government) officials and a revenue officer from Irungattukottai and Kattarampakkam, in 1998 the land acquisition officer fixed the compensation amount at Rs. 300/cent based on land prices for the 1995–8 period in the registered sale deeds (Rao 2010). Even then, many farmers did not surrender their land. Subsequently, SIPCOT served two more notices ordering landowners to surrender the land voluntarily, failing which the agency would compulsorily acquire it.

The opposition to land acquisition gained momentum in 1998. Until then, landowners continuously lobbied the two regional parties to stall the acquisition, but the opposition was not visible. In 1998 farmers from eight of the thirteen villages covered for this research came together for protest actions. They enlisted the support of elected representatives of gram panchayats, the councillor of Sriperumbudur town panchayat, lawyers in Sriperumbudur and Chingelpet, a left-leaning trade union, and a Dalit activist organization. Farmers who surrendered their land in the first round of acquisition in 1997 also joined the struggle. The landowners and their allies organized protests and public hearings and eventually filed a court case against SIPCOT. In the meantime, SIPCOT announced an enhanced compensation amount of Rs. 2000/cent for land abutting the main road and between Rs. 500–1,600/cent for the interior parcels.

Landowners' reasons for joining the struggle differed by caste, land tenure, and size of landholding. Farmers who joined the protest belonged to different caste communities, including the Naidu and Naikkar caste and the most disadvantaged Dalits. One group of farmers, predominantly from the Naidu caste community, joined the protest to demand higher compensation, especially for their wetland. SIPCOT's guidance value does not take into account the type of soil or irrigation and, consequently, the compensation offered for wet agricultural land far from the highway was lower than market value. Further, there were differences between the market price of land under agriculture and land converted into plots. Another group, predominantly small farmers from the Dalit caste, opposed land acquisition and wanted alternative land in lieu of cash compensation. They considered that the compensation amount was inadequate to invest in land in the vicinity or in alternative livelihoods.

The legal battle between landowners and SIPCOT lasted for more than a decade. The case was heard at the District Court in 1997 and at the Madras High Court between 1998 and 2010 (see Appeal Suit Nos. 742 to 758 of 2009 and connected MPs). The Madras High Court judgment ordered the Revenue Department to enhance the compensation amount to Rs. 3,750/cent and Rs. 3,500/cent after deduction of development charges (Rao 2010).

The enhanced compensation did not benefit farmers with smallholdings, ranging from less than 1 acre up to 5 acres, across different castes. According to interviews with panchayat leaders of Mambakkam and Irungattukottai villages and a DSS district-level organizer, among those owning between 5 and 7 acres only a few large landowners in each village actually secured the enhanced amount. Dalit households were relatively worse affected. The concerned government agency did not offer any compensation to a group of Dalit households dependent on common land for their livelihood. Another group of Dalit households could not secure compensation, as they held land allotted by the state under a land reform movement but had not been issued any titles.

Market Process

In contrast to Goldman's (2011a) suggestion that farmers in peri-urban locations sell their land out of fear of acquisition, findings of this

Figure 12.3 Development along the main road leading to Oragadam-Irungattokottak SEZ, India. Source: Image capture by Bhuvaneswari Raman, 10 July 2017

research show that other reasons contribute to farmers' decision to sell their land, notably declining returns from agriculture and high land prices.

Land-sale and land-use change from agriculture to urban use was prevalent in the study villages prior to SIPCOT's interventions in 1997. According to the village administrative officers of the Oragadam village, gram panchayat officials, and village-level land brokers, sale of agricultural land to real estate developers peaked between 1998 and 2000 in several villages. A combination of factors influenced farmers' decisions to sell land during this period. Farmers in some villages sold their land fearing the possibility of future acquisition by SIPCOT, but also with the desire to benefit from higher land prices in the villages around the SEZs. Others sold it for different reasons, including decline in income from agriculture, difficulty in finding labour, and maintaining tenancy cultivation. Although the declining returns from agriculture are a countrywide phenomenon, they have been particularly acute in Tamil Nadu (Vijayabaskar 2010, 2013). Interviews with farmers in different villages suggest that when SIPCOT offered Rs. 300/cent in 1997, real estate firms offered Rs. 800–1,000/cent in some villages. Land prices escalated here from 1990 onward, owing to the demand from small industrialists, who offered Rs. 1,000/cent for dry land. In the mid-1990s college entrepreneurs who required large contiguous land parcels paid up to Rs. 2,500 /cent. Prices rose further in 2004 to Rs. 6,000–8,000 for a cent of dry land. The price quoted for agricultural dry land in 2013 was around Rs. 11,000 /cent. During our field research we found that agricultural land as far as 15 km off the main highway was fallow, and some was sold out. Large farmers negotiated their sales in Chennai and these transactions are not easily visible. Very often, the tenant farmers cultivating such land were not aware of the transactions until they were concluded.

Conclusion

The production of land for suburban real estate markets is a very complex process shaped by a variety of actors and institutions embedded in different geographical and political scales. As suggested by Speculative Urbanism theory, the production of land for suburban real estate markets is driven by the logic of speculative gains from increasing land prices. The theory highlights the role of urban development authority and large developers. Besides these actors, other types of government agencies and different categories of land assemblers, developers, and land owners also are involved in the transformation process.

I want to highlight four aspects of the micropolitics of land transformation in suburban areas. One, government agencies at the regional and the local scale influence the process of transforming agricultural land for real estate markets in the suburban areas. The transformation process has two aspects, namely, land assembly and land development. Regional government agencies, by virtue of their role in regulating land-use conversion and land development, wield greater power over the process, while local governments have limited influence. Further, there are differences between regional government agencies in terms of their power and influence over land development. Besides government agencies, the courts play an important role in shaping the contours of the conflict. Two, private developers are not a unified category. Large developers collectively are one of the private actors shaping the physical landscape in the Sriperumbudur-Oragadam region. Small and medium developers also influence the process by virtue of their roles as land assemblers and developers of residential layouts with minimal construction component. Three, the transformation process is articulated by local politics and forces from outside the locality. Spatial and social networks of landowners and local political actors influence the ability to assemble land for large development projects. Large developers' ties and connections to actors in the suburban localities vary. They access land for their large project either through state allocation or purchase already assembled land. They engage with the state actors through their connections in regional politics. Those with weak or no ties in the locality rely on the regional state to access land, often through state allocation, owing to the difficulty involved in land assembly. The entry of large developers does not necessarily dis-embed small and medium developers from the region, but rather re-embeds them in new roles. Similarly, the internal contradictions of large development projects destabilize their scope for future action. The production of land is thus shaped by an interplay of culture, economy, and politics. Four, different actors adjust their roles in response to the changes in the macro-environment. While large developers' projects have slowed down, the plans of small developers have continued to proliferate. Similarly, large developers retrofit their products to enter into different retail markets with the stagnation of large development projects. Landowners respond differently to state intervention in land and the market process, depending on the perceived and actual threat of land acquisition by the industrial parastatal and the opportunities available to benefit from rising land prices.

Our findings point to the importance of a grounded reading of urban transformation and the making of urban real estate markets. Further research is required to map the reorganization of power between parastatals and local governments and how this impacts local politics. Similarly, there is a need for understanding the role of political parties, whose role in land negotiations is significant in India.

REFERENCES

Baken, R.J. 2003. *Plotting, Squatting, Public Purpose and Politics: Land Market Development, Low Income Housing and Public Interventions in India*. Aldershot, UK: Ashgate.

Balakrishnan, S. 2013. "Land Conflicts and Cooperatives along Pune's Highways: Managing India's Agrarian to Urban Transition." PhD dissertation, Harvard University.

Benjamin, S., and B. Raman. 2011. "Illegible Claims, Legal Titles, and the Worlding of Bangalore." *Revue Tiers Monde* 206 (2): 37–54. https://doi.org/10.3917/rtm.206.0037.

Chatterjee, T. 2013. "The Micro-Politics of Transformation in the Context of Globalization." *Journal of Southeast Asian Studies* 36 (2): 273–87.

Datta, A. 2015. "New Urban Utopias of Postcolonial India: 'Entrepreneurial Urbanization' in Dholera Smart City, Gujarat." *Dialogues in Human Geography* 5 (1): 3–22. https://doi.org/10.1177/2043820614565748.

Dutta, V. 2012. "Land Use Dynamics and Peri-urban Growth Characteristics: Reflections on Master Plan and Urban Suitability from a Sprawling North Indian City." *Environment & Urbanization Asia* 3 (2): 277–301. https://doi.org/10.1177/0975425312473226.

Ernst and Young. 2014. *Real Estate: Making India*. India: EY and FICCI.

Goldman, M. 2011a. "Speculative Urbanism and the Making of Next World City." *International Journal of Urban and Regional Research* 35 (3): 555–81. https://doi.org/10.1111/j.1468-2427.2010.01001.x.

Goldman, M. 2011b. "Speculating on the Next World City." In *Worlding Cities: Asian Experiments and the Art of Being Global*, ed. A. Roy and A. Ong, 229–58. Chichester, UK: Wiley-Blackwell. https://doi.org/10.1002/9781444346800.ch9.

Halbert, L., and H. Rouanet. 2014. "Filtering Risk Away: Global Finance Capital, Transcalar Territorial Networks and the (Un)Making of City Regions: An Analysis of Business Property Development in Bangalore, India." *Regional Studies* 48 (3): 471–84. https://doi.org/10.1080/00343404.2013.779658.

JLL. 2014. *Manufacturing Destinations of India: A Real Estate Overview. Property Consultants*. India: Jones LaSalle Long Pvt. Ltd.

Kamath, Raghavendra. 2014. "HDFC Puts Hirco's Chennai Township on The Block to Recover Dues." *Business Standard*. 30 January. Accessed 11 January 2017. http://www.business-standard.com/article/printer-friendly-version?article_id=114012900503_1.

Kennedy, L., et al. 2014. *Engaging with Sustainability Issues in Metropolitan Chennai. City Report 5, Chance2Sustain*. Bonn: EADI Kaiser-Friedrich-Strasse. http://horizon.documentation.ird.fr/exl-doc/pleins_textes/divers14-09/010063029.pdf.

Nandy, M. 2012. "Inno Group to Raise New Fund for Chennai Project." *Live Mint*. 2 September. Accessed 11 January 2017. http://www.livemint.com/Industry/N9WLgMbY7PxiDStX9T8lsN/Inno-Group-to-raise-new-fund-for-Chennai-project.html.

Rao. 2010: "General: When Acquired Lands Have Potential for Future Development and Situated Near Developed Area, 20% Deduction from Market Value of Land towards Development Charges Is Reasonable." *Indian Valuer* 42 (October): 1202–14.

Reuters. 2007. "Dubai's ETA Star Seeking $300mn Loan." *Arabian Business.com*, 18 September. Accessed 11 January 2017. http://www.arabianbusiness.com/dubai-s-eta-star-seeking-300mn-loan-55218.html.

Rouanet, H., and L. Halbert. 2016 "Leveraging Finance Capital: Urban Change and Self-Empowerment of Real Estate Developers in India." *Urban Studies* 53 (7): 1401–23 (pub. on line before print: 29 May 2015, https://doi.org/10.1177/0042098015585917).

Roy, A. 2014. "RBI Concerned about Banks' Exposure to Real Estate, Infrastructure." *Live Mint*. 9 October. Accessed 11 January 2017. http://www.livemint.com/Industry/tjzmLnGNYCY0FpAjfBq38K/RBI-concerned-about-banks-exposure-to-real-estate-infrastr.html.

Sami, N. 2013. "From Farming to Development: Urban Coalitions in Pune, India." *International Journal of Urban and Regional Research* 37 (1): 151–64. https://doi.org/10.1111/j.1468-2427.2012.01142.x.

Searle, L.G. 2014. "Conflict and Commensuration: Contested Market Making in India's Private Real Estate Development Sector." *International Journal of Urban and Regional Research* 38 (1): 60–79. https://doi.org/10.1111/1468-2427.12042.

Searle, L.G. 2010. "Making Space for Capital: The Production of Global Landscapes in Contemporary India." PhD dissertation, University of Pennsylvania.

Sharma, Rohan. 2013. "Integrated Townships in India – Today and Tomorrow." *The Hindu*. 20 April. Accessed 2 May 2017. http://www.thehindu.com/todays-paper/tp-features/tp-propertyplus/integrated-townships-in-india-today-and-tomorrow/article4635323.ece.

Shatkin, G. 2011. "Planning Privatopolis: Representation and Contestation in the Development of Urban Integrated Mega-Projects." In *Worlding Cities: Asian Experiments and the Art of Being Global*, ed. A. Roy and A. Ong, 77–97.

Chichester, UK: Wiley-Blackwell. https://doi.org/10.1002/9781444346800.ch3.

Thangamani, T.P. 2014. *Industries Department Policy Note 2014–2015, Demand No. 27*. Chennai: Government of Tamil Nadu.

Vijayabaskar, M. 2010). "Saving Agricultural Labour from Agriculture': Special Economic Zones and the Politics of Silence in Tamil Nadu." *Economic and Political Weekly* 45 (6): 36–43.

Vijayabaskar, M. 2013. "Tamilnadu: The Politics of Silence." In *Power, Policy and Protest: Politics of India's Special Economic Zones*, ed. R. Jenkins, L. Kennedy, and P. Mukhopadhyaya, 304–31. New Delhi: Oxford University Press.

chapter thirteen

Conclusion: What Are the Suburban Land Questions?

RICHARD HARRIS AND UTE LEHRER

Faced with the bewildering variety of suburban environments, of forms of land tenure, and of land-development processes worldwide, it is tempting to throw up one's hands in despair. How is it possible to compare or to generalize? For many decades researchers solved the problem by dividing the world into two or more regions: developed and underdeveloped; First, Second, and Third Worlds; and most recently, north and south. Labels aside, all of these distinctions were gross simplifications, which, as we argued in the introduction, worked to obscure commonalities. In particular, they directed attention away from the peculiarly transitional quality of suburban territory, transitional in geographical, cultural, and historical terms. Accepting that the land market in such areas has some distinctive characteristics, what is it that we most need to know – what questions should we ask? Although much more research has been done in some places than in others, in general we are remarkably ignorant about the land question in suburban areas, and so we believe that it is time to identify a broadly similar research agenda almost everywhere.

The most basic thing we need to know is who owns and who controls land at the urban periphery, two questions that are not necessarily providing the same answer. It determines almost everything else: Who is most likely to profit from development? What sorts of pressures will be brought to bear on the public agencies that tax and regulate land? When and where will development happen? What form will it take? There are two subsidiary issues here: how concentrated ownership is, and what types of people or companies own the land. Concentration matters because it affects the ability of owners to set prices. That is why, for example, in Canada during the 1970s there was widespread discussion about the ability of a handful of land developers to determine the price of land and housing in each urban area, prompting a national enquiry (Spurr 1976). Behind this discussion was a related concern about the ability of these developers to shape policies over land taxation and use. Concentration, in particular the size of individual holdings, can also have a significant impact on the pattern of urban development. In a general way it seems that fragmented ownership in suburban Tokyo (Plate 6a) has contributed to poor planning and a variegated landscape, while more consolidated ownership in North America (Plate 1b) has produced many large-scale planned subdivisions (Leinberger 2009; Sorensen 2001; Weiss 1987). But since precise comparisons have not been made, more nuanced conclusions are not possible.

The identity of the owners matters, too, primarily because it is likely to be associated with differing motives. Few owners are expected to be indifferent to the potential market value of their property, but for investors this is the dominant or sole motivation. Having little or no emotional commitment to the land, they may be more eager to sell. The exception is if they, or their backers, have deep pockets and are committed to a long-term time horizon. As urbanization proceeds and land prices rise, the attitudes of all of these groups may change, although probably not at the same rate. It is not enough, therefore, simply to know who the owners are; we also need to think about their motives and how suburban growth might affect them.

Undeniably, we are astonishingly ignorant about all of these matters and that is true for all parts of the typical metropolitan area. Almost half a century ago, speaking about the situation in the United States, Marion Clawson (1971, 102) commented that "of all the groups involved in the process of suburban growth, least is known about the landowners, speculators, and dealers – and considering how little we know about some other groups that is a strong statement." Since then, the lack of evidence on landownership has been lamented many times and in many different places (e.g., Evers 1976, 69; Goodchild and Munton 1985; Trivelli 1986, 120). With occasionally surprising exceptions – notably the research on Indonesian and Malaysian towns undertaken by Hans-Dieter Evers, who has devoted much of his career to the subject – we simply do not have answers to the most basic questions about the identity, concentration, and motives of property owners in urban areas (Evers and Korff 2000, 179; cf. Brown, Phillips, and Roberts 1981; Mendiola 1983).

It is not clear whether our prevailing ignorance is because few researchers and public agencies have thought to address the question or because they have found the challenge too large. Today, in most parts of the global north the task should be quite straightforward. Systems of land registration are well maintained, ownership is in principle a matter of public record, and most land records are in digital form and therefore readily susceptible to analysis. But accessing those records is rarely easy. In Ontario, for example, a private agency is responsible for gathering and updating data on property ownership and appraised values for every municipality in the province. It does so annually and provides the results to municipalities, but not readily to researchers or, except on a very limited basis, to local residents. How often researchers have attempted to gain access to such data is unclear; at any rate, to our knowledge no researcher in Ontario, or indeed anywhere in North America, has employed these records to document changing patterns of urban or suburban landownership. The same is true in those parts of the world where the challenges of undertaking such an analysis are far greater. Accurate, up-to-date records of landownership are often unavailable and, where maintained, not accessible. Obtaining information and disentangling claims for even one site can be a challenge; undertaking a broader analysis of ownership patterns has been almost impossible. Making such an effort on as large a scale as possible is, in our opinion, a high priority for research on suburban, or indeed urban, land.

Part of the task of determining ownership lies in being clear about what ownership or control means. Around the world there are often competing claims on the same sites, in part because multiple property systems coexist side by side or, what is more complex, pertain to the same property (Baken 2003). These systems may have to be reconciled, a process that should involve the people affected rather than being legislated by the state. On a smaller scale ambiguities also can arise where developers introduce private regulations and mixed-tenure arrangements in Common Interest Developments or condominium developments, and the long-term implications of this form of tenure have not been widely thought through. Knowing the prevalence and location of property that is held under various forms of land tenure, then, is one of the things we need to know if we are to make sense of suburban landownership.

It is clear enough that suburban land conversion prompts a rapid increase in prices, which sometimes happens overnight when land regulations are changed. The probability and size of a potential increase affects the mix and motives of owners, together with the likelihood that speculators and outside investors will become involved. Apart from

Figure 13.1 Land development on the urban fringe. Photo: Roger Keil

market considerations of supply and demand, the magnitude of the increase depends on how land is regulated. Many customary forms of tenure embody limitations on the ways in which land can be used, and by whom. As long as these limitations remain effective, price increases are likely to be limited (if the land, in fact, has been commodified at all). But the pressure on land that uses customary forms of tenure to fully integrate it into the market economy is driven by capitalist interests and over time is facilitated by regulations introduced by the state.

Governments, of course, regulate land use in a variety of ways, the most familiar method in North America and Europe being zoning. Planning documents that include comprehensive zoning maps may appear definitive in their attempt to demarcate and fix patterns of land use, and in the short run they can indeed shape the landscape. Any research project needs to establish their existence and acknowledge their force. But over a period of decades, and sometimes much less, such plans are flexible. Grandfathering of existing uses, exemptions of particular sites, and negotiated arrangements for many more are the norm, particularly for places with more economic and political power. Negotiations can serve various ends. In recent years a common purpose has been to achieve social goals, such as affordable housing or public green space, by allowing a developer to exceed Floor Area Ratio or other density constraints. Those same goals, and others besides, also can be achieved if the government shares in the profits of redevelopment and then directs that revenue – ideally to public purposes, not into the pockets of strategically placed officials and politicians. Profit-sharing itself can be managed in various ways, through taxation of land transfers and capital gains or through development charges. So we always need to know exactly how flexible current arrangements are, how that flexibility is managed, how money changes hands, and to what ends.

Calculating the stakes involved in suburban development involves more than a simple comparison of current prices under rural and under urban use because this gap depends on current calculations about the future. These estimates in turn hinge on expectations of continued urban growth, for there is always an anticipated increase in the value of rural property. How large this increase is depends on a variety of considerations. Some are definite: for example, the value of farmland in the Lower Mainland of Canada's British Columbia is predictably going to be affected by its limited extent, hemmed in as it is by the presence of the Pacific Ocean and the Coastal Range. Some factors are much less predictable. Recent events demonstrate that the value of urban sites in the adjacent Vancouver metropolitan area have been greatly influenced by the influx of Chinese immigrants and of investment capital, as well as by the generally strong condition of the provincial economy. At the same time, as long as urban growth continues, the provincial government will come under increasing pressure to release some of the land within what is currently the protected Agricultural Land Reserve. That is exactly the sort of pressure that London's Green Belt has been experiencing, and the outcome depends on uncertain political decisions. So the shadow increase in the price of exurban land may begin months or even decades ahead of development, as landowners and investors speculate and balance various probabilities. To understand their calculations we cannot avoid coming to terms with the range of possible near- and medium-term outcomes.

A large part of the challenge in making sense of land-development processes at the urban fringe is that of figuring out who actually services and controls the land. It would be wrong to assume that governments always do so. In many parts of Latin America, Africa, India, and some parts of Europe and North America basic infrastructure is provided by pirate land developers. In both the global north and the global south, developers view such infrastructure provision as one of the necessary costs of doing business. The difference is that in one region they voluntarily make these investments and are able to exercise wide discretion

Figure 13.2 Land development on the urban fringe of Pittsburgh, USA. Photo: Ute Lehrer

about the range and quality of the services they provide, while in the other they are required to install specific services and to meet minimum standards. Except in Common Interest Developments, they are also expected to eventually turn over the ownership, as well as responsibility for maintenance, to public authorities (though with neoliberalization we find more and more privatized services such as water, electricity, and even road maintenance). The significant questions, then, concern the presence and nature of standards and the allocation of costs between public and private agencies.

Who has control over land development is in many ways the most significant question, and the largest conundrum. The fundamental challenge is that of distinguishing between nominal and practical powers. Legislation usually defines the responsibilities of government and the extent of their powers. From such laws the services that public agencies are supposed to provide, and the powers of land-use regulation and taxation they possess, should be clear. But what they actually do is usually another matter. If they fail to provide and especially to maintain roads, sewers, and so forth, are they held to account and if so how? Do they actually enforce the regulations that are on the books and if not why and with what consequences?

It may seem that governments fail to enforce regulations for two types of reason: because they are unable to, or because they choose not to (Harris 2017; Roy 2005, 2015). In fact, although helpful in some ways, this distinction is overly neat. Most fundamentally, if municipalities neglect to implement all of their regulations, it is because they either lack the resources to do so or have other priorities. The deficiency is a worldwide phenomenon and ubiquitous. No municipality in the world is able to proactively inspect and enforce all of its building and land-use regulations or its building maintenance, health, and occupancy by-laws. Politicians and municipal staff know this and allocate their scarce resources accordingly. Although they rarely announce the fact, they turn a blind eye to infractions, a practice for which the Dutch even have a name: *gedogen*. Again, this is particularly common for people and in areas with economic and political power. In many cases this practice is managed on an ad hoc basis and may not even rise to the level of a conscious decision. In other instances at least parts of it may reflect strategic decisions about where best to allocate limited funds and human resources. Sometimes permission for informal practices is monetized through payoffs by property owners, whether to low-level inspectors or to their supervisors and political masters. The important question, therefore, is not whether regulations are enforced, but how. We need to know which ones are enforced and to what degree; how infractions are managed, by whom, and why.

In one sense, it can be argued that infractions, or instances of informality, lie on a continuum. A complicating factor has been the argument that as informality grows, it crosses thresholds, so that it is helpful to think of distinctive modes (Harris 2017). Regardless, probably one of the most important challenges for those who are interested in the land question in suburban areas is to figure out how regulation and informality are managed. Certainly, it is likely to be the most difficult task. Municipalities do not like to advertise their informal practices or failures, except in limited ways and on specific occasions when they are seeking approval from their electorates for an increase in taxes. Making public a decision not to enforce certain by-laws is likely to encourage more builders, developers, or property owners to flout the law. Advertising the fact that inspectors respond to bribes will invite condemnation officially, but unofficially it is common that people, agencies, and corporations involved in the specific developments are working exactly on that assumption. Therefore, it is better to put the word around unobtrusively. For various reasons, then, and predictably, in almost all situations it will be very difficult to determine how the world of suburban land development actually works, except through close ethnographic research.

The challenges are large: establishing who owns suburban land and on what terms; what those owners or later developers stand to gain from its development; and what role governments play in servicing and regulating that land, not just in principle but also, and more important, in practice. That is not the end of it. As we argued in the introduction, these issues are best viewed historically, above all because suburban environments are transitional between the countryside and the city. This aspect is most obviously true in terms of their physical appearance – land-use mix, density – and market value. In many countries it is also true of the tenurial form and legal status of the land. No snapshot, however accurate and precise, can capture the unpredictable dynamic that emerges from the interplay of those elements. But for all the challenges it is worth trying. The stakes are high, above all for the residents of suburban environments, who are not a negligible group. In recent years the media have made much of the fact that the world is now urban. What we need to add is that, because so much of that urbanization has happened recently and is proceeding so rapidly, a large proportion of the urban world is, in fact, suburban in the ways that we have discussed. Attention should be paid.

REFERENCES

Baken, R.J. 2003. *Plotting, Squatting, Public Purpose and Politics: Land Market Development, Low Income Housing and Public Interventions in India*. Aldershot, UK: Ashgate.

Brown, H.J., R.S. Phillips, and N.A. Roberts. 1981. "Land Markets at the Urban Fringe: New Insights for Policy Makers." *Journal of the American Planning Association* 47 (2): 131–44. https://doi.org/10.1080/01944368108977098.

Clawson, M. 1971. *Suburban Land Conversion in the United States*. Baltimore: Johns Hopkins University Press.

Evers, H.-D. 1976. "Urban Expansion and Landownership in Underdeveloped Societies." In *The City in Comparative Perspective*, ed. J. Walton and L.H. Masotti, 67–79. New York: Sage.

Evers, H.-D., and R. Korff. 2000. *Southeast Asian Urbanism: The Meaning and Power of Social Space*. New York: St Martin's Press.
Goodchild, R., and R. Munton. 1985. *Development and the Landowners: An Analysis of the British Experience*. Boston: Allen & Unwin.
Harris, R. 2017. "Modes of Informal Urban Development: A Global Phenomenon." *Journal of Planning Literature*. https://doi.org/10.1177/0885412217737340.
Leinberger, C. 2009. *The Option of Urbanism: Investing in a New American Dream*. Washington, DC: Island Press.
Mendiola, E.C. 1983. "Urban Land Reform in the Philippines." In *Land for Housing the Poor*, ed. S. Angel, R. Archer, S. Tanphiphat, and E.A. Wegelin, 473–500. Singapore: Selecte Books.
Roy, A. 2005. "Urban Informality: Towards an Epistemology of Planning." *Journal of the American Planning Association* 71 (2): 147–58. https://doi.org/10.1080/01944360508976689.
Roy, A. 2015. "The Land Question." LSE Cities. https://lsecities.net/media/objects/articles/the-land-question/en-gb/. Accessed 15 August 2017.
Sorensen, A. 2001. "Building Suburbs in Japan: Continuous Unplanned Change on the Urban Fringe." *Town Planning Review* 72 (3): 247–73.
Spurr, P. 1976. *Land and Urban Development: A Preliminary Study*. Toronto: Lorimer.
Trivelli, P. 1986. "Access to Land by the Urban Poor." *Land Use Policy* 3 (2): 101–21. https://doi.org/10.1016/0264-8377(86)90048-7.
Weiss, M. 1987. *The Rise of the Community Builders: The American Real Estate Industry and Urban Land Planning*. New York: Columbia University Press.

Contributors

Wolfgang Andexlinger, Faculty of Architecture, University Innsbruck & City of Innsbruck, Austria. Wolfgang Andexlinger is Associate Professor at the Faculty of Architecture at the University Innsbruck, Austria. Since August 2016 he also has been head of the Department for Urban Planning, Urban Development and Integration at the City of Innsbruck, Austria. In research and teaching his thematic focus lies in the consideration of urbanization processes and their spatial implications. He holds a PhD and a Habilitation in Urban Design and Regional Planning. In 2011 he was visiting scholar at the City Institute (CITY) at York University in Toronto, Canada. His most important publications include *TirolCITY – New Urbanity in the Alps* (2005); *Superimpositionen – Neue Stadt-Landschaften in den Alpen am Beispiel von Sölden* (2012); *Alpine Stadt-Landschaften – Beobachtungen zur Rolle von Freiräumen im Tiroler Inntal aus Sicht des Landscape Urbanism* (2013); *Suburban Processes of Islandisation in Austria: The Cases of Vienna and Tyrol* (2015), and *Tirol im Kontext globaler Urbanisierungsprozesse* (2016).

Pierre Filion, School of Planning, University of Waterloo. Pierre Filion is Professor at the University of Waterloo School of Planning. His research interests include downtown and inner-city planning, metropolitan regional planning, and land-use–transportation interaction. He is currently working with the Major Collaborative Research Initiative on *Global Suburbanisms: Governance, Land and Infrastructure in the 21st Century*, particularly the sections on infrastructures and transportation. He is also leading a research project focusing on recentralization as a metropolitan planning strategy in Canada's largest metropolitan regions. He has published 132 refereed journal articles and book chapters and has written and co-edited fourteen books and special journal issues. Recent publications include "Recentralization as an Alternative to Urban Dispersion: Transformative Planning in a Neoliberal Context" (with A. Kramer and G. Sands), *International Journal of Urban and Regional Research*, 2017; "Contested Infrastructures: Tension, Inequity and Innovation in the Global Suburb" (with R. Keil), *Urban Policy and Research*, 2016; "Planners' Perspective on Obstacles to Sustainable Urban Development: Implications for Transformative Strategies" (with M. Lee, N. Leanage, and K. Hakull), *Planning Practice and Research*, 2015.

Alan Gilbert, Department of Geography, University College London. Alan Gilbert is Emeritus Professor at UCL. His research focuses on urbanization and poverty in Latin America and the global south. Gilbert has written on housing policy, renting, informality, poverty, inequality, social segregation, and urban governance. He has authored or co-authored nine books, including *Latin American Development: A Geographical Perspective* (Penguin Books); *Cities, Poverty and Development: Urbanization in the Third World* (with Josef Gugler; Oxford University Press); *Housing, the State and the Poor: Policy and Practice in Three Latin American Cities* (with Peter Ward) (Cambridge University Press); *Landlord and Tenant: Housing the Poor in Urban Mexico* (with Ann Varley; Routledge); *The Latin American City* (Latin America Bureau and Monthly Review Press); *Rental Housing: An Essential Option for the Urban Poor in Developing Countries*, (UN-HABITAT); and *Bogotá: progreso, gobernabilidad y pobreza* (with María Teresa Garcés; Universidad del Rosario: Bogotá). He has prepared a number of consultancy reports for the Inter-American Development Bank and several agencies of the United Nations. In 2003 he was elected to the Academy of Learned Societies for the Social Sciences.

Jill L. Grant, School of Planning, Faculty of Architecture and Planning, Dalhousie University. Jill Grant's research seeks to illuminate the relationship between theory and practice in planning. Her interests focus on the design and planning of residential environments with a special focus on trends such as new urbanism and gated communities. Recent research projects examine neighbourhood change and how planners coordinate multiple plans. Dr Grant has published over sixty academic articles and five books: *The Drama of Democracy: Contention and Dispute in Community Planning* (University of Toronto Press, 1994); *Planning the Good Community: New Urbanism in Theory and Practice* (Routledge, 2006);

Towards Sustainable Cities: East Asian, North American and European Perspectives on Managing Urban Regions (with A. Sorensen and P. Marcotullio; Ashgate, 2004); *A Reader in Canadian Planning: Linking Theory and Practice* (Thomas Nelson, 2008); and *Seeking Talent for Creative Cities* (University of Toronto Press, 2014).

Richard Harris, School of Geography and Earth Sciences, McMaster University. A Fellow of the Royal Society of Canada and of the Royal Canadian Geographical Society, Richard Harris is Professor of Geography at McMaster University, Hamilton, Canada, and currently president of the Urban History Association. He has published widely on the modern history of residential segregation, housing policy, homeownership, the housebuilding industry, and suburban development, with particular reference to North America, Kenya, and India. His most recent books are *Building a Market: The Rise of the Home Improvement Industry, 1914–1960* (University of Chicago Press, 2012), and *What's in a Name? Talking about Urban Peripheries* (co-edited with Charlotte Vorms; University of Toronto Press, 2017). He is currently writing a book about the history of neighbourhoods in Canadian cities since 1900.

Sonia Hirt, School of Architecture, Planning and Preservation, University of Maryland. Sonia Hirt is Dean and Professor at the School of Architecture, Planning and Preservation at the University of Maryland, College Park. Her focus as a scholar and teacher is on exploring the complex social meanings of the urban built environment. She aspires to help enhance the quality of urban environments first, by developing a richer understanding of the social processes and cultural values that influence their evolution, and second, by provoking critical debates within the urban planning and design professions. Previously, Hirt served as Professor and Associate Dean at Virginia Tech and Visiting Associate Professor at the Graduate School of Design, Harvard University. She is the author (with K. Stanilov) of *Twenty Years of Transition: The Evolution of Urban Planning in Eastern Europe and the Former Soviet Union, 1989–2009* (UN-HABITAT, 2009); *Iron Curtains: Gates, Suburbs and Privatization of Space in the Post-Socialist City* (Wiley-Blackwell, 2012) and *Zoned in the USA: The Origins and Implications of American Land Use Regulation* (Cornell University Press, 2015). Her research has been funded by organizations such as the American Council of Learned Societies, the American Association of University Women, and the National Endowment for the Humanities.

Françoise Jarrige, UMR Innovation, Montpellier SupAgro, INRA, CIRAD, Univ Montpellier, Montpellier France. Françoise Jarrige is an economist of public policies and innovation processes in the fields of land, agriculture, and territories. She specializes in the issues of farming and farmland in urban policies and projects. She teaches social sciences, with a focus on land and natural resources management, in major education programs at Montpellier SupAgro. She has contributed to several French and European research projects on periurban agriculture.

Pia Kronberger-Nabielek, Centre of Regional Planning and Regional Development, Technical University of Vienna. Pia Kronberger-Nabielek is a researcher and lecturer in the Department of Spatial Planning at the Technical University of Vienna. Her research work has an international focus and concentrates on institutional and spatial challenges in European regional planning concerning urban growth, energy transition, and climate adaptation. Furthermore, she explores participative methods in scenario-oriented research. Currently, for her PhD, she is researching wind-power deployment in highly urbanized regions in Austria, Belgium, and the Netherlands. She has authored or coauthored the books *TirolCITY: New Urbanity in the Alps* (2005) and *Flood Risk as a Spatial Planning Issue* (2006). More recent publications concern the report *Adaptive Strategies for the Rotterdam Unembanked Area* (2013) and the articles "Flood Risk Data and Urban Design" (2014), and "Balanced Renewable Energy Scenarios" (2017).

Ute Lehrer, Faculty of Environmental Studies, York University. Ute Lehrer is a member of the City Institute and an Associate Professor in the Faculty of Environmental Studies, York University. She holds a PhD in Urban and Regional Planning from the University of California, Los Angeles and a lic. phil in Art and Architectural History, Sociology, as well as Economic and Social History from the University of Zurich, Switzerland. Previously, she taught at the Department of Geography, Brock University and at the School of Architecture and Planning, SUNY Buffalo, and was a researcher at the Swiss Federal University of Technology in Zurich. She also worked as an architecture critic for the largest daily newspaper in Switzerland. Lehrer has published widely on urban form and economic restructuring, globalization and built environment, architecture, and urban planning. She currently serves on the editorial board of the *International Journal for Urban and Regional*

Research and was the founding editor of the peer-reviewed journal *Critical Planning*. She has held various positions on international committees, among them ISA-RC 21, where she was the vice-president for North America. Lehrer is a founding member of INURA (International Network for Urban Research and Action). She is a co-applicant of the Major Collaborative Research Initiative "Global Suburbanisms" and the Principal Investigator for a comparative study on Frankfurt and Toronto funded by the Social Sciences and Humanities Research Council of Canada.

Zhigang Li, School of Urban Design, Wuhan University. Zhigang Li is Professor of Urban Studies and Planning at the School of Urban Design, Wuhan University, one of the top universities in mainland China. Since 2015 he has served as Dean of the school, and previously he worked for a decade in Sun Yat-sen University, Guangzhou, China. He received his PhD in Urban Geography from the University of Southampton in the UK in 2005. Li's research focuses on the social spatial issues of Chinese cities. His recent work concentrates upon the impacts of globalization and market reform on urban China's socio-spatial structures. He has conducted many studies in the communities of Guangzhou, Dongguan, Foshan, Shenzhen, and Wuhan to examine new migrant suburbs such as "chengzhongcun." Li has published over 100 papers in both Chinese and English, in a number of top journals such as *Transactions of the Institute of British Geographers*, *Urban Studies*, and *Antipode*. He is a co-applicant of the "Global Suburbanisms" Major Collaborative Research Initiative.

Kersten Nabielek, PBL Netherlands Environmental Assessment Agency. Kersten Nabielek works as a researcher and designer at the urban department of the PBL Netherlands Environmental Assessment Agency. His research interests focus on urban and suburban developments in Dutch cities, the relationship between urban mobility and spatial planning, and comparing spatial, demographic, environmental, and socio-economic developments in European cities. Recent publications include "The Compact City: Planning Strategies, Recent Developments and Future Prospects in the Netherlands" (2012) and "The Rural-Urban Fringe in the Netherlands" (2013). Furthermore, he has authored the books *Cities in the Netherlands* (2016) and *Cities in Europe* (2016). Currently, he is working on future scenarios of urbanization and new forms of mobility in the Netherlands.

Emmanuel Négrier, Centre d'Études Politiques de l'Europe Latine, CNRS-University of Montpellier. Emmanuel Négrier is CNRS Senior Research Fellow at Montpellier, in political science. He is a member of the Scientific Commission of the INRS (Canada), and editor of Pôle Sud, a French journal of political science. His main fields of investigation are political rescaling and territorial reforms, cultural policies and political behaviours. He collaborates in several cultural policy groups or organizations, including Observatoire des politiques culturelles (Grenoble), Fondation de France (Paris), University of Barcelona (Cultural Policies Area), Département des études et de la prospective du ministère de la Culture (Paris), as well as Creative Europe (Brussels) and France Festivals (Paris). He is a collaborator on the "Global Suburbanisms" Major Collaborative Research Initiative.

Bhuvaneswari Raman, Jindal School of Government and Public Policy, Jindal Global University. Bhuvaneswari Raman's research focuses on urbanization, governance, and poverty in south Asian countries. She has researched and written on urban transformation, real estate dynamics, and the politics of urban development, informality, poverty, and urban governance. She received her PhD in Social Policy from the London School of Economics in 2010. Her recent research focuses on transnational urbanism in India and China and land transformation in Indian metros and small towns. She has consulted for international organizations, including the World Bank, UNCHS and UNESCO. She has published in national and international journals such as the *Economic and Political Weekly*, *Journal of South Asian Development*, *Tier Monde* (currently *Journal of International Development*), and *Critical Planning*.

Margot Rubin, South African Research Chair in Spatial Analysis and City Planning in the School of Architecture and Planning and a Research Associate with the Society, Work and Development Institute (SWOP), both located at the University of the Witwatersrand, Johannesburg. Since 2002 she has worked as a researcher and policy and development consultant, focusing on housing and urban development issues, and has contributed to a number of research reports on behalf of the National Department of Housing, the Johannesburg Development Agency, SRK Engineering, World Bank, Ekurhuleni Metropolitan Municipality, and Urban LandMark. Her PhD in Urban Planning and Politics interrogates the role of the legal system in urban governance in India and South Africa and its effect on the distribution of scarce resources and larger questions around democracy. She also holds a Masters degree in Urban Geography from the University of Pretoria, an Honours degree in Geography and Environmental Studies, and a Bachelor of Arts in Geography and Philosophy. In her work as the Research Chair, Margot has been writing about inner-city regeneration and housing policy and is currently engaged in work around mega housing projects and issues of gender and the city.

Christian Schmid, Professor for Sociology, Department of Architecture, ETH Zürich. Christian Schmid is a geographer and sociologist. Since 1980 he has been a video activist, an organizer of cultural events, and an urban researcher. In 1991 he was co-founder of the International Network for Urban Research and Action (INURA). Schmid has authored, co-authored, and co-edited numerous publications on urban development in the Zurich region, theories of the urban and of space, Henri Lefebvre, territorial urban development, and the comparative analysis of urbanization. With architects Roger Diener, Jacques Herzog, Marcel Meili, and Pierre de Meuron he co-authored the book *Switzerland: An Urban Portrait*, a pioneering analysis of extended urbanization. Currently, he is working with Neil Brenner on the theorization and investigation of emergent formations of planetary urbanization and leading a project on the comparison of urbanization processes in Tokyo, Pearl River Delta, Kolkata, Istanbul, Lagos, Paris, Mexico City, and Los Angeles, which is based at the ETH Future Cities Laboratory Singapore. He is a collaborator on the "Global Suburbanisms" Collaborative Research Initiative.

Marc Smyrl, Centre d'Etude Politique de l'Europe Latine (CEPEL), Université de Montpellier. Marc Smyrl is Associate Professor of Political Science at the University of Montpellier and research fellow of the CEPEL. His current research centres on the comparative dynamics of policy change in the United States and the EU with a particular focus on the health and technology sectors. Recent publications include "Beyond Interests and Institutions: US Health Policy and the Surprising Silence of Big Business," *Journal of Health Politics, Policy, and Law* (2014) and "Innovation ouverte et 'Living Labs': Production et traduction d'un modèle Européen," *Revue Française d'administration publique* (2017).

André Sorensen, Department of Human Geography, University of Toronto Scarborough. André Sorensen is Professor of Urban Geography at the University of Toronto. His current research examines planning, urban property, and temporal processes in urbanization and urban governance from an institutionalist perspective. His paper "Taking Path Dependence Seriously" (2015) (*Planning Perspectives* 30 (1): 17–38) won the Association of European Schools of Planning Best Paper Award in 2016.

Fulong Wu, Bartlett Professor of Planning, University College London. Fulong Wu's research interests include urban development in China and its social and sustainable challenges. He has an extensive record of publications, including being co-editor (with Laurence Ma) of *Restructuring the Chinese City* (Routledge, 2005), editor of *Globalization and the Chinese City* (Routledge, 2006), editor of *China's Emerging Cities: The Making of New Urbanism* (Routledge, 2007), co-author (with Jiang Xu and Anthony Gar-On Yeh) of *Urban Development in Post-Reform China: State, Market, and Space* (Routledge, 2007), co-author (with Chris Webster, Shenjing He, and Yuting Liu) of *Urban Poverty in China* (Edward Elgar, 2010), co-editor (with Chris Webster) of *Marginalization in Urban China* (Palgrave Macmillan, 2010), and co-editor (with Nick Phelps) of *Comparative Suburbanism* (Palgrave Macmillan, 2010). His most recent book is *Planning for Growth: Urban and Regional Planning in China* (Routledge, 2015). He is a co-applicant of the Major Collaborative Research Initiative "Global Suburbanisms."

Index

Note: (t) = tables; (f) = figures

GLOBAL SUBURBANISMS

Series Editor: Roger Keil, York University

Published to date:

Suburban Governance: A Global View / Edited by Pierre Hamel and Roger Keil (2015)

What's in a Name? Talking About Urban Peripheries / Edited by Richard Harris and Charlotte Vorms (2017)

Old Europe, New Suburbanization? Governance, Land, and Infrastructure in European Suburbanization / Edited by Nicholas A. Phelps (2017)

The Suburban Land Question: A Global Survey / Edited by Richard Harris and Ute Lehrer (2018)